Modern
Grain Sorghum
Production

William F. Bennett
Billy B. Tucker
A. Bruce Maunder

Iowa State University Press / Ames

Modern Grain Sorghum Production

William F. Bennett is Associate Dean and Professor of Agronomy, College of Agricultural Sciences, Texas Tech University.

Billy B. Tucker is Regents Professor of Agronomy, Agronomy Extension Leader, and Extension Agronomist, Emeritus, at Oklahoma State University.

A. Bruce Maunder is Vice-President, Agronomic Research, at DEKALB-PFIZER GENETICS, Inc., and currently serves as chairman of the US/AID Title 12 International Sorghum Millet Collaborative Research Support Program External Evaluation Panel.

Manufactured in the United States of America

♾ This book is printed on acid-free paper

First edition, 1990

Library of Congress Cataloging-in-Publication Data

Bennett, William F.
    Modern grain sorghum production / William F. Bennett, Billy B. Tucker, A. Bruce Maunder. — 1st ed.
        p. cm.
    Includes bibliographical references.

    ISBN 0-8138-1121-X
    1. Sorghum. 2. Sorghum industry. I. Tucker, B. (Billy), 1928–   . II. Maunder, A. Bruce. 1934–   . III. Title.
SB191.S7B46   1990
633.1'74—dc20   89-13470

CONTENTS

v

PREFACE

GRAIN SORGHUM has historically been one of the five major cereal crops used for food by humankind. Its origin can be traced to Africa and India. It is one of the oldest crops known to humankind. In addition to use as food, grain sorghum is also used as feed for animals. The sorghum plant has many different shapes, forms, and sizes. Starting with very tall, low-yielding sorghums grown in Africa, plant breeders and geneticists have been able to manipulate the plant to the point that it can be a high-yielding, high-quality grain for use both as food and as a feed.

This book is designed for use by anyone interested in or related to grain sorghum production—the producers; the agribusiness industry that provides seed, chemicals, fertilizer, machinery, and other production needs; those who handle the grain including brokers and processors; and technical and advisory groups including consultants, teachers, extension workers, ag loan bank lenders, and federal agency personnel such as ASCS, SCS, FHA, BIA, and managers of wildlife preserves.

The book can also be used as a text for high school vocational agriculture courses. Student and teacher guides are available for this use.

A few readings and individual references are given, but because of its general nature, we have not attempted to give specific references, since it is from the general body of knowledge that information was drawn.

The authors wish to express appreciation to our friends and professional co-workers who assisted with ideas and thoughts on this book and gave us encouragement to prepare it.

We also express appreciation to the many researchers who have developed information on grain sorghum. It is these individuals who provide the specific information that becomes the general body of scientific knowledge on which a book of this nature is based.

It is not possible to recognize the contributions of all the researchers who have added to the storehouse of knowledge on grain sorghum. In many cases, the original author of certain specific facts is not known. We acknowledge the contributions of all researchers who have worked to improve this important crop.

We wish to acknowledge individuals, companies, and corporations who furnished slides and pictures for the book. They include Dale Hollingsworth, Leon New, Jack Musick, Stanley Coppock, Irvin Williams, Dale Weibel, and Stanley Osle as individuals. Organizations and foundations include the Potash-Phosphate Institute and the High Plains Research Foundation. Companies and corporations include DeKalb/Pfizer Genetics, Gifford-Hill Western, Goodpasture, John Deere, Tennessee Corp., Tyler Industry, Kan-Sun, and Butler Co.

Modern
Grain Sorghum
Production

1

Introduction

THE MODERN GRAIN SORGHUM PLANT (*Sorghum bicolor* L. Moench) is a product of human ingenuity. This marvelous energy-producing plant has been selected, domesticated, and changed to fit human needs. It has proven enormously useful in areas too hot and too dry to grow good yields of corn. In recent years, it has been adapted to many parts of the world.

Origin

Early records show that sorghum existed in India in the first century A.D. A carving depicting sorghum was found in an Assyrian ruin dating from 700 B.C. However, sorghum probably originated in Africa, although some evidence indicates that it may have had independent origins in both Africa and India. The wild types found in central East Africa and India are unsuitable for use in today's agriculture, but plant breeders continue to search out these wild types looking for new germplasm in order to incorporate desirable characteristics into present genetic lines.

Sorghum as a domesticated crop found its way to Europe about 60 A.D. but was never grown extensively on that continent. It is not known when the plant was first introduced into the Americas. The first seed was probably brought to the Western Hemisphere in slave ships from Africa. The U.S. Department of Agriculture (USDA) released the first sorghum seed in 1857. Since that time, numerous sorghum materials have been introduced from many parts of the world by the USDA and other agencies. The durras, both white and brown, were brought to California from Egypt in 1874. Kafirs and other classes of sorghum were introduced into the United States soon thereafter. Shallu found its way from India about 1890, kafirs

3

from South Africa about 1904, and feterita and hegari from the Sudan a few years later.

The first sorghum crop of record in the United States was grown by William R. Prince of New York in 1853.

The early sorghums left a lot to be desired as a grain crop. They were very tall, making them susceptible to lodging and difficult to harvest. In addition, they were very late to mature and were successful only in the southern regions of the United States. Kafir and milo types were selected to use as grain by the early settlers in the Great Plains because of their greater tolerance to drought than corn. With the advent of machine harvesting, selections were made from the original materials that gave earlier and somewhat shorter types. However, it was the development of combine-type grain sorghums, pioneered by John B. Sieglinger of Oklahoma, that made it possible to grow sorghums using mechanized harvesting. The further development of the early types as well as disease- and insect-resistant varieties together with other improved production practices firmly established grain sorghum as an important grain crop throughout the central and southern Great Plains and in other portions of the western United States by 1940 (Fig. 1.1).

The most dramatic progress, however, was yet to come. As a result of the research by J. R. Quinby and J. C. Stephens of Texas,

1.1. *An excellent field of grain sorghum.*

hybrids became a reality by 1960. Yield levels up to 12,000 pounds per acre are now being grown. Such yields are possible because of new hybrids with high-yield potential and the use of production practices such as adequate fertilization; weed, disease, and insect control; and high plant populations.

Adaptation

Grain sorghums are generally cultivated in areas that are too dry or too hot for successful corn production. They originated in the tropics but are now adapted as far north and south as latitudes of 45 degrees. They are cultivated extensively in Africa, India, Manchuria, Argentina, and the United States. Some sorghum is also grown in other parts of Asia, Europe, Central America, and South America.

Sorghums are adapted to the drier climates because of several factors. They have the ability to remain dormant during drought and then resume growth. Sorghum leaves roll as they wilt; thereby less leaf surface area is exposed for transpiration (Fig. 1.2). The leaves and stalks of sorghum contain an abundance of waxy covering, thereby protecting it from drying. Sorghum plants exhibit a low transpiration ratio (pounds of water required to produce a pound of

1.2. *Grain sorghum during moisture stress showing rolled leaves. (Photo by J. T. Musick, USDA-ARS)*

plant material). In Table 1.1, relative transpiration ratios are given comparing sorghum with other important field crops.

Table 1.1. Transpiration ratios for sorghum and selected crops

Crop	Pounds water/lb of plant tissue dry matter
Sorghum	311
Corn	375
Sugar beets	397
Wheat	532
Potatoes	554
Alfalfa	904

Note: Data are from Colorado. The same approximate differences have been shown in other experiments.

Sorghums have a large number of fibrous roots that effectively extract moisture from the soil (Fig. 1.3). It has been estimated that the absorption area of the root system of a sorghum plant is twice that of corn. This large absorption capacity and a relatively small leaf area are other factors in its drought resistance.

1.3. *Grain sorghum has a large, extensive root system. (Courtesy De-Kalb)*

In addition to the tendency for growth resumption when relieved of moisture stress, the sorghum plant also produces new tillers whenever moisture becomes available if drought has not been prolonged.

Even though sorghums will produce good yields with high temperatures, best yields are obtained when mean temperatures during the growing season are in the 75–80° F range. The mean midseason temperature should normally exceed 70° F. High daytime temperatures normal for most areas are not detrimental provided nights are cool and soil moisture is adequate. Sorghums are able to withstand temperatures over 100° F, but dry winds coupled with hot weather during pollination will reduce yields.

Another reason sorghums are adapted to areas with warmer climates is that they require a given number of heat units for certain phases of development. Head initiation and exertion are controlled by daylength. Sorghums will not usually head when daylight exceeds 16 hours. They cannot be grown to produce high yields too much farther north of 45° latitude north or south of 45° latitude south.

It is fortunate that soil characteristics in the hot and dry climates normally lend themselves to good sorghum production (Fig. 1.4). Sorghums can tolerate a wide range of soil pH and textures. Soils in the lower rainfall areas have inherently higher soil pH values than those in the more humid areas.

1.4. *This excellent field of grain sorghum produced a high yield under dryland conditions.*

Grain sorghums are grown mostly where annual rainfall ranges from 15 to 25 inches or where irrigation water is available to supplement rainfall. The growing season normally should be longer than 130 days, even though very early maturing varieties mature with a shorter season.

Acreages

The world acreage of sorghum is also increasing rapidly. Asia is the leading continent in sorghum acreages, as shown in Table 1.2. In Asia, India is the largest producer. Asia is followed by Africa in sorghum acreage, where the leading sorghum countries in Africa are Nigeria, Ethiopia, Niger, and Mali, but sorghum is produced to some extent in almost every African country. In Central and North America, the United States (Fig. 1.5), Mexico, and Haiti produce the bulk of the crop. In South America, Argentina is the principal producer.

Table 1.2. World sorghum acreages

Continent	Acres × 1000
Asia	49,864
Africa	26,064
North America	12,817
Central America	2,026
Australia	452
South America	403
Europe	237

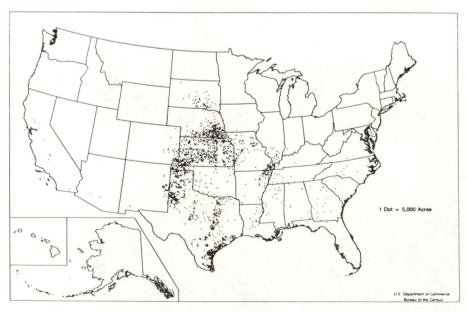

1 Dot = 5,000 Acres

U S Department of Commerce
Bureau of the Census

1.5. Grain sorghum is grown in the United States primarily in the Great Plains, where Texas, Kansas, and Nebraska are the leading producers.

Sorghum Names

Sorghum is classified into four distinct groups: (1) grain or non-saccharine sorghum, (2) forage or saccharine sorghum (sorgo), (3) broom corn, and (4) grass sorghum.

The first group includes the grain sorghums with old types such as kafir, milo, and feterita. It also includes the dwarf combine sorghums and the combine-type hybrids. The second group includes those grown principally for forage, silage, or molasses. The third group is a special group in which the panicle (or head) is used for brooms. The fourth group is used chiefly for hay and pasture.

Because of the many kinds and types of sorghum, a host of names has come into common use. Sorghum is the general name for the crop. It is similar to crop names like wheat, corn, and barley. Both grain and forages are a part of the sorghum crop. Included in sorghum are groups that include milo, kafir, feterita, and sorgo. Names often heard include Dwarf Yellow Milo, White Kafir, Red Kafir, Feterita, Hegari, and Dorso.

As crosses of these groups were made and as more and more varieties and hybrids were developed, additional names coming into use included maize, milo maize, hybrid milo, hybrid maize, kafir corn, and milo. These names are usually used interchangeably by farmers. Some of the varieties that were developed and persisted until hybrids became a reality included Plainsman, Martin, Midland, and Westland.

The sorghum used for the harvest of grain are now properly called grain sorghum. The term maize is often applied to grain sorghum, but it refers more properly to corn because of its Latin name, *Zea mays*.

2

The Grain Sorghum Plant

The modern grain sorghum plant is only 20–50 inches tall as contrasted to the types grown before the 1940s (Fig. 2.1). The taller sorghums have been replaced with "double dwarf" or combine types to facilitate mechanical harvesting.

There are at least seven grain sorghum groups that have been introduced into the United States. These broad groups are referred to as kafir, milo, feterita, durra, shallu, kaoliang, and hegari. The most common of the groups grown commercially are kafir, milo, feterita, and hegari.

2.1. *Normal short-grain sorghum hybrids are shown in the foreground; a taller type of forage sorghum can be seen in the background. (Courtesy Phosphate Potash Institute)*

10

The kafirs were introduced in the Americas from South Africa and are identified by their thick, large, juicy stalks; large leaves; and white, pink, or red seed.

The milos from central East Asia have wavy leaf blades with a yellow midrib. The heads are compact with an oval shape and the glumes are dark brown. The seed are large and of a salmon or creamy white color. It was the introduction of milo types with their high degree of heat and drought tolerance that caused sorghums to spread into the drier climates. The early milos had a curved peduncle and were commonly called "crooked-neck maize."

The feteritas were introduced from the Sudan. They have slender stalks and few leaves. The heads are oval and compact. The seed are very large and white in color.

Hegari resembles kafir but differs in shape of the heads, which are more oval than kafir. The seed of hegari are chalky white. Hegari plants tiller profusely. In earlier days they were known as dual-purpose sorghums because they produced considerable grain on a rather large plant and were popular for "bundle" feed harvested for both grain and forage.

The milos and kafirs were the most used types in the development of modern varieties and hybrids. Present-day hybrids cannot be adequately classified into any of the sorghum groups listed above. They may resemble a group in some characteristics but are not fully typical of any one group. It is for this reason that common usage of the term "milo" to denote modern grain sorghum is a misnomer. The same is true of the term "maize" used in some localities.

Several of the taller types of grain sorghums (Fig. 2.2) are still produced in many countries where mechanized harvesting is not important and production of both forage and grain is needed. Some of the varieties cultivated in Africa are 8 feet or more in height.

2.2. Hybrids vary in height because of differences in length of internodes. The stalk of grain sorghum has 7–12 nodes and internodes. A leaf is borne at each node, alternating on each side of the stalk. All of these hybrids have a similar genetic makeup except for the gene that controls height.

The aboveground culms of sorghum have 7–12 nodes and internodes. The internodes are longest and most uniform in the middle of the stalk and shortest at the base, but the terminal internode that bears the head is often the longest of all. A leaf is borne at each node. Basal diameters of the culms vary from 0.25 to 2 inches, depending primarily upon plant population. The mature stem is quite pithy; some are sweet but others are not. The nodes alternate on opposite sides of the culm. The surface of the culms, sheaths, and leaves is glaucous.

Some sorghum varieties tiller profusely under favorable environmental conditions. Hegari and feterita types tiller more profusely than the milos. Tillering ability of modern hybrids depends upon the parentage. Tillering is usually undesirable in production of grain because it often causes uneven ripening and thus may create problems at harvest.

The number of leaves varies from 7 to 18 or more. The length of the vegetative period determines the number of leaves. Each additional leaf adds 3–4 days to the length of the growing season. The greater number of leaves and, consequently, the greater leaf area, probably account for the highest yields being obtained from the longer season hybrids.

All the leaves of the plant are developed during the first 3–4 weeks of the plant's development. New leaves are produced by a single growing point at the top of the stem. New leaves appear until shortly before booting, but all these leaves were formed inside the plant during the vegetative period of growth.

The first leaf below the head is known as the flag leaf; this is the leaf whose sheath holds the top of the boot prior to head exsertion (Fig. 2.3).

At the base of each leaf is a long sheath, which overlaps in dwarf varieties. The leaf margins have sharp, curved indentations. Numerous motor cells are located near the midrib on the upper side of the leaf. These cells facilitate rapid folding of the blade during periods of moisture stress.

Sorghum leaves contain a large number of stomata. It has been estimated that sorghum leaves possess 50% greater numbers of stomata per unit area than corn, but the stomata are smaller. The leaves are coated with a heavy bloom of white wax, which helps limit evaporation so that the leaves wilt more slowly than those of corn.

The sorghum head or inflorescence is a panicle (Fig. 2.4). It ranges from 3 to 20 inches in length and from 2 to 8 inches in width. Some are classified as open heads, with long primary branches on a hairy axis. The seed branches are borne in whorls or clusters. Compact head types have short primary branches. The fertile sessile spikelet consists of a short floral axis. The seed are covered by two glumes, referred to as the first and second glume. The glumes are somewhat indurate and about equal in length. They are usually

2.3. *The mature sorghum head ranges from 3 to 20 inches in length, with the stem on which it is carried called a peduncle and the first leaf below the head called the flag leaf.*

2.4. *Dissected grain sorghum head showing the seed clusters. The seed are borne in whorls or clusters. Compact head types have short primary seed branches. (Courtesy DeKalb)*

black, red, brown, or straw colored. The glumes enclose two florets; the upper one is fertile while the lower one is normally sterile. The fertile floret is made up of the lemma and palea, two lodicules, three stamens, and a one-celled ovary with a bifurcated plumose stigma. Both lemma and palea are hyaline and delicate. On awned hybrids and varieties, the awn is attached to the lemma of the fertile floret. Most often the sterile pedicellate spikelets fall off soon after the fertile spikelets mature.

The peduncle carrying the head is ordinarily straight and about 1–6 inches long. Sometimes it becomes curved or kinked because of resistance to the pressure of head emergence from the sheath. Some varieties have peduncles that remain green and succulent even after grain maturity. This helps to keep the head erect for harvesting and reduces head loss. Peduncle thicknesses vary from 0.1 to 1 inch in diameter and bear a rather close relation with culm diameters of the plant.

The seed or caryopsis of grain sorghum vary in size from about 0.05 to 0.1 inch. Seed color is quite variable and includes white, chalky white, pink, yellow, red, buff, brown, and reddish brown. Some of the white-seeded varieties are spotted with red, purple, or brown pigments that wash from the glumes.

In most modern sorghum varieties and hybrids, there are from 12,000 to 28,000 seed in a pound. An average of 15,000 per pound is generally used for present hybrids when estimating quantity of seed to plant. The kernel shape is usually obovoid.

The grain sorghum kernel is made up of three major parts: the pericarp or the seed coat, the endosperm or starch material, and the embryo or germ. The pericarp protects the seed, while the starch is used for energy and food in sustaining life of the germ. It is the germ or embryo that gives life to the new plant. The genetic makeup of the plant is in the embryo. All ingredients necessary for reproduction are present along with oil and many substances that are active and important to sustain life in the very early stages of germination and growth. The sorghum kernel consists of about 84% endosperm, 10% germ, and 6% pericarp. The endosperm starch is composed of both straight-chain and branched-chain starches. Some varieties produce seed containing carotene and xanthophyll in the endosperm.

Sorghum has a fibrous root system like most grass crops. The first root develops from the radicle of the seed. From this single seminal root, lateral roots develop along its length. The seminal root may function during the entire life of the plant but ceases to be of importance after the permanent root system begins to function.

The permanent roots are adventitious and develop in succession from the basal nodes near the ground level. The whorls of adventitious roots in their entirety constitute the root crown. These "brace" roots continue to grow quite profusely until the booting stage. From the permanent roots, lateral roots are highly branched, interlacing in all directions, especially near the surface. From the 8- to 10-inch depth the number of lateral roots decreases.

A third set of roots develops just above the permanent roots on nodes above the ground level. These "aerial" or buttress roots are considerably thicker than normal roots and are usually green in color. On entering the soil the aerial roots behave as ordinary roots and decrease in size to the diameter of the main laterals.

Sorghum roots can penetrate to depths of 5 feet or more, but the majority are in the upper 25 inches of soil, with a heavy concentration in the top 10 inches. Lateral roots grow to as much as 10 feet to the side of the plant in the soil surface and are found abundantly from 20 to 30 inches away from the plant by the time it has reached the 4- to 6-leaf stage.

Compared to corn, sorghum has more secondary roots, giving it the ability to extract more of the available soil water. It has been estimated that the absorption capacity of the root system of a sorghum plant is twice that of corn. Sorghum roots are finer and more fibrous than those of corn.

Physiology

The grain sorghum plant is unique in its ability to produce energy so efficiently and under conditions too adverse for most grain crops. It can make some growth and reproduce on low-moisture and low-nutrient levels but responds very well to improved conditions and good management.

GERMINATION

When the sorghum seed is placed in a warm, moist environment, several changes begin to take place (Fig. 2.5A). First, water enters the seed coat and the kernel swells. This activates the chemical reactions that initiate growth. The first sign of new growth is when the radicle (root) elongates and emerges from the kernel (Fig. 2.5B). Shortly thereafter, the plumule also begins to elongate and

2.5.A. *Sorghum seed should be placed in a warm, moist environment for proper germination. (Courtesy DeKalb)*

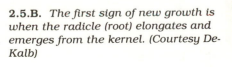

2.5.B. *The first sign of new growth is when the radicle (root) elongates and emerges from the kernel. (Courtesy DeKalb)*

2.5.C. *The radicle continues to elongate while the plumule (shoot) begins to emerge from the top of the seed. (Courtesy DeKalb)*

2.5.D. *The mesocotyl (shoot) grows upward and breaks through the soil surface. (Courtesy DeKalb)*

additional leaves form in the cotyledons (Fig. 2.5C). The radicle grows downward and is anchored into the soil; at the same time the mesocotyl grows toward the soil surface, the seed (cotyledon) emerging with it (Fig. 2.5D). Lateral roots form along the length of the seminal root (Fig. 2.5E).

The amount of time required after planting for the stem to emerge varies depending primarily upon temperature, moisture, and seeding depth, but under favorable conditions it requires from 5 to 10 days. The radicles have elongated and anchored prior to stem emergence.

The seed should be placed into soil with adequate but not excessive moisture, and the soil temperature should exceed 65°F. If the soil is too wet, too dry, or too cold, germination will be slow. If germination is too slow, the seedling is subject to attack by detrimental soil organisms. The coleoptile is pointed and protected by hard cell tips. If soil compaction is too great, the growing tip cells may rupture below the ground, in which case it would not be able to push its way out; after twisting and turning it often dies. Layers of hard soil are often caused by too much compaction of packer wheels or surface crusting following high-intensity rainfall.

Sorghum seed should be covered with from 1 to 2 inches of soil. Deeper plantings can be successful under ideal conditions, but under unfavorable situations, the coleoptile will become weakened if it must traverse through too much soil. The mesocotyl will ordinarily elongate about 1 inch and the coleoptile must lengthen the rest of the distance to the soil surface in order to bring the stem the rest of the way aboveground.

VEGETATIVE GROWTH

After germination, the sorghum plant goes through a vegetative growth period (Fig. 2.6). During this time, the plant is becoming

2.5.E. *As the mesocotyl reaches the soil surface, lateral roots form along the length of the seminal root. (Courtesy DeKalb)*

prepared for the final stage of fruiting or maturity; roots and leaves are developing rapidly.

The stalk during this period becomes larger as each new node with leaves appears (Fig. 2.7). The leaves are produced from the growing point, which is underground or near the ground, the first 3 weeks after planting. The boot begins to form as soon as all leaves have emerged.

2.6. *With a satisfactory environment the young seedling emerges and begins its vegetative growth. (Courtesy DeKalb)*

2.7. *The stalk is enlarging as each new node with leaves appears. At the 4-leaf stage the growing tip is still below-ground, but the head is already beginning to form.*

It requires only 2 weeks after emergence for four leaves to appear. It is at about the 4-leaf stage that the head begins to form, but it is formed belowground in the growing tip. By the time six leaves have fully developed, the head is aboveground and beginning to enlarge. The boot becomes visible by the 6- to 8-leaf stage and emerges after all leaves are formed (Fig. 2.8). The flag leaf, whose sheath holds the head, is the last leaf to appear.

2.8. *By the time six leaves have fully developed, the head is aboveground and beginning to enlarge.*

During the later stages of vegetative growth, the sorghum plant must be operating at full efficiency if maximum yields are to be obtained. Moisture and nutrients are being rapidly absorbed during this period. However, the grain sorghum plant throughout the vegetative growth period can undergo stress and recover. If drought occurs, leaves roll and the plant becomes semidormant. Photosynthesis and other activities are slowed until conditions are more favorable.

REPRODUCTIVE GROWTH

As the plant begins the fruiting process, vegetative growth starts to slow. The plant at this stage puts all its resources into the reproduction process and the production of grain. Water and nutrient absorption is reaching its peak, and adverse conditions at this time can have a drastic effect on yield.

The developing head is held in the upper whorl or growing tip (Fig. 2.9). As the head develops, the upper leaf swells. The enlarged leaf whorl enclosing the panicle is called the boot. As the head expands, the upper leaf allows the head to slide out. The head exerts through the sheath of the upper leaf. This upper leaf is referred to as the flag leaf. After exertion, the upper internode (peduncle) elongates, raising the head above the flag leaf (Fig. 2.10).

In a plant like sorghum with a terminal inflorescence, duration of growth and ultimate size are dependent upon the time when floral

2.9. *The boot (head) becomes visible by the 8- to 10-leaf stage and the head emerges after all the leaves are formed.*

2.10. *As the head expands in the boot, the upper leaf allows the head to slide out. After complete exsertion, the upper internode elongates, raising the head above the flag leaf.*

initiation takes place. Head initiation can take place anytime the vegetative growth requirements are fulfilled, providing the day-length is not too long. Sorghum as a short-day species, and maturity is hastened by shorter days. Recently, sorghum cultivars have been developed for more tropical locations, which are less sensitive to day-length.

In most varieties the flowers do not begin to open until the head is completely out of the sheath, but only a day or two is required after the head is exserted. The first flowers open near the apex of the panicle and progress downward to the base at the rate of 1–2 inches per day (Fig. 2.11). The anthers (source of pollen) come loose and burst open as soon as the glumes are spread (Fig. 2.12). The bloom-ing process of a single spikelet takes only a short time, often less than an hour. The flowers normally remain open for a few hours, but if fertilization does not take place, the flowers remain open for

2.11. *Flowering, starting at the top of the head and progressing downward, begins soon after head exsertion from the boot. The blooming process in a head takes from 4 to 15 days.*

2.12. *By this stage, all of the stigma should have been fertilized.*

several days. Flowering usually takes place at night or early morning. This is probably due to fluctuating temperatures between night and day, because in cool weather blooming often takes place as late as midmorning. The blooming of a single panicle takes from 4 to 15 days. The cooler the temperature the longer it takes.

Pollen deteriorates rapidly under adverse conditions. During the flowering period, hot dry winds can be particularly damaging. If the pollen is dehydrated too rapidly, "blasting" of the heads occurs, leaving the head with unfilled grain. Low soil moisture at this time can also be quite damaging.

The stigmas seem to remain receptive to pollination for several days, even up to 2 weeks, after the flowers bloom. Fertilization usually takes place within 6–12 hours after pollination. After fertilization, the newly formed embryo cell is inactive for about 4 hours. Organ differentiation in the embryo begins on the seventh day and is completed on the twelfth day (Fig. 2.13). The embryo continues to grow until the seed is mature.

2.13. *The reproductive part of grain sorghum—a perfect flower— showing both staminate (male) and pistillate (female) parts.*

Sorghums are self-pollinated. The pollen produced in each floret normally fertilizes the stigmas on that floret and surrounding florets on the head. Cross-pollination occurs easily whenever foreign pollen is available. When sorghum varieties are grown in adjacent rows, the amount of cross-pollination averages from 4 to 8%. Wind is the chief source of pollen transport.

The number of pollen available for fertilization of the stigma is great. There are more than 5000 pollen grains per anther in most hybrids and varieties. This gives more than 20 million pollen grains per panicle. The large number of pollen grains is evident from those covering the plants and ground during flowering.

MATURATION AND GRAIN DEVELOPMENT

In the discussion on germination, it was noted that the seed contains three parts: the germ or embryo, the endosperm or food reserve, and the seed coat or pericarp. It is the embryo that will become the new plant if the seed is planted. As soon as fertilization takes place, the seed begins to develop. Apparently for the first few hours after fertilization, no measurable reactions take place. Organ differentiation begins on the seventh day after fertilization and is completed on the twelfth day.

Pollen not used for fertilization drops, and most pollen is gone from the head in 15 days. In about a week after fertilization, small watery blisters can be noted at the end of the stigmas. By the end of the second or third week after pollination, the kernels begin to take shape and are filled with a milky substance. This is the "soft-dough" stage, when the kernel is high in sugars. It is during this stage that bird damage can be most severe, because the kernel is highly palatable in most hybrids and varieties. Some varieties have been developed that contain bitter tannins, which will later decline as the grain ripens. These varieties are valuable where bird damage is especially severe and for planting in wildlife refuges. Bird damage is usually severe when small fields of sorghums are interspersed with pasturelands, forest, or plantings of nongrain crops.

As the kernel becomes more mature, the water content decreases and the sugars give way to gummy dextrins and finally drier starch. This is the "hard-dough" stage, which under normal conditions occurs about the fourth week after flowering.

Total quantities of nitrogen, total sugars, starch, and acid-hydrolyzable carbohydrates are greatest when the grain has reached maximum dry weight, which occurs from 4 to 6 weeks after pollination. After this period, the dry weight decreases as a result of continued respiration activity. By the time the seed dry to a moisture content suitable for conventional harvesting and storage, some dry weight yield is lost. Maximum yields are obtained several days in advance of usual harvest dates, which might suggest either early harvesting and the use of artificial drying or finding methods of storing wet grain.

Maximum nitrogen percentages are found in the grain early in its development. Nitrogen percentages decrease until 18–24 days after pollination and remain relatively constant thereafter. The sugar content in young developing kernels is mainly sucrose. It reaches its maximum percentage at about 8 days after pollination. The sugars are transposed into more complex forms of carbohydrates. The starch content makes up about 60% of the total dry matter of the kernel near maturity.

During kernel development, adverse environmental conditions will reduce kernel fill and determine whether the lower part of the head will have kernels at all. It is at this stage that yields can either be surprisingly high or disappointingly low. It is kernel size that is primarily affected.

All changes taking place after maximum dry weight is reached are concerned with drying and harvesting. The moisture content must drop to about 14% for conventional combine harvesting and storage. At this moisture content the grain becomes inactive.

Kernels dry from top to bottom on the head. The panicle and peduncle dry to varying degrees depending upon the variety or hybrid. Varieties with panicles and peduncles that remain green are subject to less grain loss than those without this "stay green" characteristic.

Rainfall after maturity and drying results in grain weathering. The extent of damage depends on the amount of rainfall, humidity, temperature, and wind. If it is too damp for long periods, mold and algae growth may deteriorate the grain. Excessive moisture for long periods may even result in grain sprouting in the head. Heavy rain after maturity can cause discoloration of the kernel. Pigments can be washed from the glumes to discolor the kernel. This causes no damage and does not affect quality. Rain after maturity may also bleach the grain, making it lighter in color. Bleaching of grain is of no real detriment to quality. Excessive rainfall may cause the test weight (weight per volume) to decrease.

Plant Nutrition

The growth of a plant is accomplished by photosynthesis, which is the process whereby light energy (solar energy) plus carbon dioxide and water are converted into food energy in the form of carbohydrates. In the process, plants give off oxygen, which is vital to all animal life, including human. This is truly an amazing production process. Water enters the roots and is conducted to the leaves, where most of the photosynthesis takes place. Carbon dioxide enters through small openings in the leaf called stomata. The simple carbohydrates formed are sugars, which are subsequently converted to starch and other plant compounds.

In addition to elements required in photosynthesis (carbon, hydrogen, and oxygen), the plant requires a wide variety of other elements. These include the essential macroelements—nitrogen, phosphorus, potassium, sulfur, calcium, and magnesium—and the essential microelements—iron, zinc, manganese, copper, boron, molybdenum, and chlorine. The microelements are just as essential as the other elements but are needed only in small (micro) amounts. The microelements are often referred to as minor elements, trace elements, or micronutrients.

NITROGEN

Nitrogen is needed by the plant for structural purposes (Fig. 2.14). It is a constituent of protein, which is the chief ingredient of the cytoplasm. Therefore, there is a constant need for nitrogen throughout the life cycle of the plant. The demands for nitrogen follow closely the dry matter production of plants. In order to properly manage nitrogen applications, it is important to know the nitrogen uptake patterns by morphological stages of plant growth. By knowing these patterns, a grower can better determine the appropriate time and placement of side-dressings of supplemental nitrogen needed to supply the crop's full nitrogen requirements. In the seedling stage, only small amounts of nitrogen are needed; however, some nitrogen in the early stages is critical.

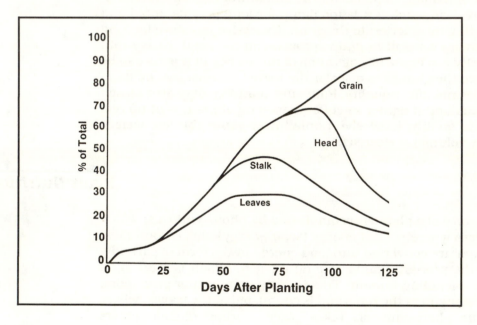

2.14. *Nitrogen uptake by the plant is continuous to physiological maturity, decreasing around midseason in the leaves, stalks, and head, and moves into the grain.*

PHOSPHORUS

Phosphorus has been called the "maturity" element because plants deficient in it are later in maturity. Phosphorus is involved in one way or another in nearly all metabolic processes in the plant. It is essential for cell division and enlargement, for photosynthesis, and for the energy transfer system of the plant. Phosphorus stimulates early root growth, helps seedling vigor, and is important for seed production.

Even though phosphorus is classified as a macroelement, it is used by grain sorghum in relatively small quantities as compared to nitrogen and potassium; but this does not diminish its importance to the growth of the sorghum plant.

Phosphorus needs are relatively large during the early growth stage. A large percentage is taken up during the plant's early growth and is stored in the leaves. It is readily translocated to the grain as the head develops. In the early stages of grain development, the ratio of phosphorus in the stalk and grain is about 1:1, whereas late in grain development the ratio changes to about 1:2. By maturity, a 6000-pound-per-acre grain yield will contain approximately 12 pounds of phosphorus, but the stubble that produces it will contain about 7 pounds. The phosphorus content of leaves in young sorghums average about 0.6% and drops to about 0.3% near maturity.

POTASSIUM

Potassium plays many roles in plant growth. One important function is in carbohydrate synthesis. It also acts as an enzyme activator and plays a key role in regulating water relationships within a plant, keeping the osmotic pressure of the cell sap properly regulated and adjusting stomatal movement and resulting water vapor movement or transpiration. Potassium is unique in that it does not enter into the structural makeup of the plant, nor is it a component of any plant compound. It causes plants to be more succulent, helps in disease and insect resistance, and reduces stalk breakage. Potassium uptake in sorghums is most rapid during the younger stage. Potassium concentrations are higher in the vegetative tissue than in the grain. Leaf potassium averages about 2% in the boot stage but drops during flowering and is again high during late grain development, often averaging 2.5–3%. A 3000-pound grain sorghum yield will contain about 20 pounds of potassium, whereas the stubble producing the grain contains about 110 pounds.

OTHER NUTRIENTS

The rest of the essential elements and their functions in the plant are listed in Table 2.1

Table 2.1. **Essential plant nutrients other than N, P, and K and their function in plant growth**

Plant nutrient	Function in plant
Calcium	Forms calcium pectate used in cell walls; enzyme activator
Magnesium	Central atom of the chlorophyll molecule; energy transfer
Sulfur	Essential part of three amino acids essential for protein formation
Iron	Component of chlorophyll; cofactor for enzymatic reactions; electron transfer
Zinc	Involved in auxin metablism; part of dehydrogenase enzyme; enzyme activator
Manganese	Electron transport and part of enzyme systems
Copper	Part of oxidase enzyme systems
Molybdenum	Part of nitrate reductase enzyme
Boron	Growth and development of new meristematic cells
Chlorine	Osmotic and cation neutralization

Plant Composition

Grain sorghum generally fits the same pattern of uptake and distribution of the essential elements as for other grain crops. However, differences in the ability of grain sorghum compared to corn to absorb nutrients have been noted by differences in the chemical composition of the leaves of these plants. The grain sorghum plant is usually higher in nitrogen, phosphorus, and zinc than the corn plant grown under the same conditions. Corn is usually higher in potassium and iron than grain sorghum. Fertilizer will influence nutrient uptake and chemical composition of the plant.

The chemical composition of any plant will vary according to conditions under which it is grown. Critical levels have not been definitely established for grain sorghum; however, guidelines have been established for ranges that can be expected under normal conditions. Table 2.2 gives some ranges that have been reported for the first leaf below the flag leaf of the grain sorghum plant. Time of sampling is preboot to early boot stage. Ranges are given for the chemical composition of grain also.

Table 2.2. Normal ranges in composition of the leaf and grain of grain sorghum

Element	Range		Element	Range	
	Leaf	Grain		Leaf	Grain
	(%)			(ppm)	
Nitrogen	2.0–3.0	1.0–2.0	Iron	15–150	40–60
Phosphorus	0.2–0.4	0.2–0.4	Zinc	10–75	5–15
Potassium	1.5–3.0	0.3–0.5	Manganese	10–100	5–15
Calcium	0.3–0.5	0.03–0.05	Copper	5–10	5–15
Magnesium	0.2–0.4	0.1–0.2	Boron	10–75	1–3

Water Requirement and Drought Resistance

As has already been noted, the grain sorghum plant has the remarkable ability to withstand drought. It was mentioned that several factors were responsible for this resistance: (1) a large, efficient root system, (2) a relatively small transpiration capacity in relation to the large root absorption capacity, (3) the ability to roll its leaves and cut down on transpiration during moisture stress, and (4) the waxy covering on leaves and stems, which protects it from excess water loss.

Grain sorghum needs from 15 to 25 inches of usable soil moisture for maximum production. It has already been shown that it uses water more efficiently than most crops. It can extract moisture to several feet (in most soils 4–6 feet) but uses water efficiently to only 30 inches. Highest yields are obtained when moisture is available in the top 30 inches of soil. The major zone for moisture extraction is about 30 inches for irrigated sorghum compared to a depth of over 7–8 feet for alfalfa.

Water is used by grain sorghum beginning with seed germination. At the seedling stage the amount of moisture taken up is small; therefore, soil moisture supplies are not rapidly depleted, but it is critical that sufficient soil moisture be present. As the leaves develop and transpiration increases, water use rises rapidly and reaches a peak at the boot stage. At this peak water demand period, sorghums use more than 3 inches per day; for best yield, adequate moisture should be continuously available. During the vegetative growth period (prior to booting), however, water shortage is not so critical, since at this stage of growth the plant has the ability to go into physiological dormancy without a great reduction in yield, providing conditions improve in time for the crop to recover. The most critical period for adequate soil moisture is after the water use peak has been reached. During booting (head exertion) and flowering, soil moisture is critical. During this period, water requirements for sorghums are as high and as critical as they are for corn. The ability of the sorghum plant to withstand drought seems to be exhibited before head initiation. After the head emerges, shortages of soil moisture greatly reduce potential yields. Severe moisture stress during the flowering period often results in blasted heads, with reduced seed formation.

Even though the rate of water usage decreases after head exertion, ample soil moisture is necessary for good filling of the kernels. After the maximum dry matter weight of the head has been reached, soil moisture is of little consequence to the yield. By this time, plentiful supplies of water are left in the plant tissues for normal maturity.

In summary, water stress at any time after planting reduces yields. Sorghums can tolerate drought best between the 7-leaf stage and early boot (20–40 days after planting). Adequate supplies of available soil moisture (at least 0.2 inch per day) are necessary during booting and flowering (40–65 days after planting). Adequate moisture should be present through the soft-dough stage if maximum yields are to be obtained. Excessive soil moisture after the dough stage sometimes results in tillering or suckering, which delays drying and harvesting.

3

Soil Management

SOIL MANAGEMENT refers to the manipulation of soil factors affecting crop production so that optimum yields can be achieved at lowest cost. This means that the physical, chemical, and biological conditions of the soil are in a satisfactory state whereby plants will grow at an optimum rate. Residue management, cropping systems, and tillage practices are related to soil management.

Residue Management

Soil management has often been described as "the proper handling of plant residues as related to nitrogen application and need." Handling of grain sorghum stubble would be a logical place to start in discussing a grain sorghum production program.

Decomposition of stubble prior to the planting of the next crop is usually desirable. If the stubble is decomposed, there will be less interference from the stubble during tillage and planting. Early decomposition also means that plant nutrients will be available earlier the following spring. It also keeps the grain sorghum stubble from "tieing up" nitrogen as it decomposes the following spring and thereby creating a temporary nitrogen deficiency early in the life of the following crop.

Grain sorghum is often harvested early enough in the fall so that incorporated residue can be at least partially decomposed. Soil temperatures would still need to be sufficiently high for decomposition to occur. The practice of shredding the stalks and incorporating as early as possible has proven to be beneficial from this standpoint.

For decomposition to occur, microbiological activity is necessary. Conditions for this to occur include proper temperature, adequate moisture, good aeration, an adequate level of nutrients, and the proper soil reaction. Most agricultural soils contain the proper types and numbers of bacteria necessary for decomposition. If all of these conditions are right, the stubble will be decomposed. If any one of the factors is limiting, such as low soil temperature, little decomposition will take place.

Some grain sorghum fields are grazed after harvest (Fig. 3.1), or the stubble may be cut and baled and then used for roughage for cattle (Fig. 3.2). The leaves are a good source of roughage and energy. Whether to graze or to incorporate early is a question often raised by growers. Some question whether there will be sufficient returns from grazing to overcome the benefits of early incorporation. Putting a large number of cattle on the field for a short time immediately after harvest is an approach to utilizing the stubble and permitting early incorporation.

One main point of concern in grazing grain sorghum is the possibility of it containing substances that would be toxic to cattle. It is known that if sorghum is stricken by drought or is under stress due to early frost or if there is regrowth after frost, the prussic acid content may be sufficiently high to be toxic to livestock. If grain sorghum grows normally, the prussic acid content will usually be sufficiently low so that there would be no adverse effect when grazed by cattle. The main point is to be cautious in grazing the new growth that follows harvest of plants that have been under stress due to drought or an early frost. A more complete discussion on prussic acid accumulation is given in Chapter 7.

Tieing up of nitrogen due to stubble decomposition presents a problem to growers. This tie-up is caused by a high ratio of carbon to

3.1. *The stalks and leaves of grain sorghum are still partially green after harvest and provide a good source of roughage for cattle. (Courtesy High Plains Research Foundation)*

3.2. *An alternative to grazing is to bale the stubble for later use by cattle. (Courtesy Gifford-Hill Western)*

nitrogen in the residue. If carbon is more than 20 times greater than nitrogen (C:N ratio > 20), the bacteria decomposing the stubble will need some nitrogen from the soil in addition to that in the stubble. This means that less nitrogen will be available for the following crop until decomposition has occurred to the point that the C:N ratio is again at 20. If the stubble is incorporated early in the fall, the tie-up of soil nitrogen will take place in the fall when no crop is growing, resulting in a release of the nitrogen earlier the following spring.

The C:N ratio in grain sorghum stubble will vary depending on fertilizer rates used, growing conditions, and other cultural practices. It will nearly always exceed 40:1 and will normally range from 60:1 to 80:1. This means that microorganisms that decompose the residue will use nitrogen already in the soil from some source other than the residue. The only other nitrogen source is from the soil. If soil nitrogen is low, fertilizer may need to be applied.

One advantage of bacteria tieing up soil nitrogen in the fall is that the inorganic nitrate form is changed to an organic form. Nitrogen in this form is not subject to leaching. Therefore, where leaching during the winter is a problem, it would be desirable to have the nitrogen tied up by bacteria.

An application of nitrogen on grain sorghum stubble is often advised in order to hasten decomposition (Fig. 3.3). Whether additional nitrogen should be applied will depend on the amount present in the soil.

There are conditions under which decomposition will occur and where additional nitrogen applications would be beneficial. If grain sorghum stubble is incorporated into the soil when soil temperature is low, the application of nitrogen would be of little benefit in hastening decomposition since there would be minimal microbial activity. On the other hand, if soil temperatures are above 65°F and if other

3.3. *Nitrogen solutions are often sprayed on grain sorghum stubble to provide extra nitrogen for the decomposition process. (Courtesy Goodpasture, Inc.)*

conditions are proper, a nitrogen application could hasten stubble decomposition.

Various rules of thumb have been given on the amount of nitrogen to apply to stubble. A suggested rate is about 10–15 pounds per 1000 pounds of stubble produced. A grain sorghum crop will usually produce equal quantities of grain and aboveground vegetative growth. A 6000-pound-per-acre yield of grain would produce about 6000 pounds of stubble; hence, 60–90 pounds nitrogen per acre would need to be applied if it is desirable to hasten decomposition.

The use of nitrogen to aid in decomposition of unwanted seed has been practiced by some seed producers. Where sudan crosses are used, producers will shred and incorporate the stubble from these rows early and apply nitrogen to aid in the decomposition of the seed so that they will be less of a problem to the following crop.

Nitrogen:sulfur ratios in grain sorghum stubble have been used as a criterion to determine whether sulfur may be needed to aid in decomposition. The N:S ratio should be about 15.

Grain sorghum stubble is often used as an aid in controlling wind erosion and assisting in the conservation of moisture. Stubble mulching by leaving the residue partially exposed accomplishes this effect. Partial decomposition of the portion incorporated into the soil will proceed if conditions are right, leaving only the aboveground portion to be decomposed later.

Cropping Systems

Cropping systems involving grain sorghum are quite variable over the grain sorghum–producing areas. Various cropping systems or rotations are often employed for reasons such as better utilization of resources, reducing risks due to weather or prices, control of diseases or insects, herbicide residue problems, taking advantage of the residual nutrients, weed control, and improving the physical condition of the soil.

CONTINUOUS GRAIN SORGHUM

Continuous grain sorghum is quite often grown in areas where adequate moisture or irrigation water is available. Yields can be maintained at a high level provided the necessary cultural practices are adopted, including adequate fertilization, weed control, and irrigation. Proper residue management is of importance in such cases.

Continuous cropping to grain sorghum has generally shown no detrimental effects on physical condition of the soil; it may even improve it. Continued use of certain mechanical practices year after year under continuous cropping may have a detrimental effect on soil physical condition. To be sure that the physical properties are not deteriorating, watch for telltale signs that might develop, which could indicate problems and thus suggest a need to change to some other crop. Such signs might include lowered yields, slow drying of the soil, cloddy seedbeds, and land being harder to plow. Pests such as diseases, weeds, or insects may become a problem with continuous grain sorghum.

ROTATIONS

No consistent patterns of cropping systems have emerged in the grain sorghum growing areas. A popular rotation for both dryland and irrigated land in the wheat area is grain sorghum-fallow-wheat. A crop of grain sorghum is planted in the spring and harvested in the fall. The stubble is often left in the field until the following spring, when tillage begins and the land is fallowed until fall. Wheat is then planted on the fallowed land to be harvested in early summer of the following year. This permits the producer to utilize water more efficiently and spread out available irrigation water over a longer season. Such a rotation is often practiced where weeds are a problem, with the summer season being used to control weeds. This is often the case with bad infestations of johnsongrass and shattercane. It also permits growers to graze the stubble into the fall and winter. In some instances, producers will plant wheat immediately after the grain sorghum is harvested, or grain sorghum is often planted in wheat stubble.

In the warmer portion of the principal grain sorghum–producing regions where cotton is a principal crop, rotations of cotton and grain sorghum are often used (Fig. 3.4). Where cotton root rot and verticillium wilt are a problem, grain sorghum in the rotation will aid in reducing the effect of these diseases.

3.4. *Rotating crops with grain sorghum is practiced in some parts of the sorghum-producing area. Contour farming with terraces reduces water runoff and soil erosion. (Courtesy High Plains Research Foundation)*

In grain sorghum areas where vegetables and sugar beets are grown, grain sorghum will often be rotated with these crops. Such crops are usually heavily fertilized. The grain sorghum crops that follow can then benefit from the residual plant nutrients.

In areas of dryland production where moisture is usually the limiting factor (less than 15 inches per year), the cropping system may consist of 1 year of sorghum production followed by a year of fallowing the land to conserve moisture. Less of this is done, however, than for a crop like wheat or other small grains.

Intercropping of grain sorghum with other crops is practiced in a few areas; however, this practice is limited.

CROPS FOLLOWING SORGHUM

Growers for many years have noted that some crops following sorghum do not grow and yield as well as crops following other crops. There are several possible explanations for this. The very efficient root system of sorghum extracts considerable quantities of available moisture and nutrients, and unless either one or both are replenished, the following crop will suffer. Soil organisms tie up large amounts of available soil nitrogen in decomposing sorghum residue. This effect can be overcome by nitrogen fertilizer additions. Seedbeds are often more difficult to prepare following sorghums due to the large root crowns. Another possible effect is the presence of phytotoxic substances produced by the decaying residue. The limitations of crops following sorghum can be overcome by careful management practices, including application of plant nutrients, good tillage, and moisture replenishment.

Tillage and Land Preparation

Land preparation for grain sorghum is practiced in many ways. The need for and the type of tillage carried out will depend on the soil type, cropping systems and residues, weeds and other pests, and the need to conserve moisture and control erosion. Tillage is usually practiced to modify the soil, improve soil-air-moisture relationships, prepare for irrigation or leveling, control weeds, manage crop residues, apply fertilizers and chemicals, improve water movement into the soil, and/or decrease erosion due to either wind or water.

Tillage of soil may vary from farms on which there is no tillage of soil (except planting)—called "no-till" systems—to some farms where the soil may be tilled eight to ten times. Due to high costs, the trend for most producers is to reduce the number of tillage operations.

The initial tillage operation in grain sorghum production is usually to incorporate residue, kill weeds, and conserve moisture. This may be done by plowing, disking only, or stubble mulching. Types of plows commonly used are the moldboard plow, the disk plow, the chisel plow, and subsurface sweeps.

The moldboard plow lifts and turns over a slice of soil. It buries weeds and trash and aerates the soil. This type of plow is effectively used in much of the grain sorghum area. It is generally used more in areas of irrigation and adequate rainfall.

The disk plow (Fig. 3.5) is normally used where plant residue is light and where it is desirable to leave some residue on the surface as a mulch or to prevent wind erosion. The disk plow is commonly used in the dryland areas.

The subsurface-type implements usually consist of large sweeps that are pulled beneath the surface at 4–6 inches (Fig. 3.6). They lift the soil and consequently aerate and pulverize it. A disking of the land will usually precede the sweeping.

In some areas, "duck-foot" sweeps and field cultivators are used to prepare seedbeds. When this system is used, the initial tillage is usually with a chisel, followed by the small sweeps as needed to control weeds and prepare a seedbed.

The finer textured soils, such as a silty clay loam, are usually plowed earlier than sandy-textured soils. This permits the soil to "mellow" due to freezing and thawing or wetting and drying action.

"Deep-plowing" is practiced in some areas. This may be from 10–12 inches in some areas (usually every year or every other year) to 20–30 inches in other areas. This latter type of deep-plowing is usually done only once every 10–20 years for the purpose of mixing a light-textured surface and subsurface soil with a deeper, finer tex-

3.5. *Incorporating residue with a disk plow. (Courtesy John Deere)*

3.6. *A practice often used is to plow with a subsurface sweep (stubble mulching) and apply ammonia at the same time. Note that sorghum residue is left on the surface. (Photo by Dale Hollingsworth, Texas Tech University)*

tured clay or clay loam material. This provides a more desirable soil texture in the upper soil horizons that is less vulnerable to wind erosion and has a better water-holding capacity and a higher level of plant nutrients (Fig. 3.7). Deep-plowing should be carefully considered and the subsurface soils checked thoroughly before such a practice is undertaken. One aspect of deep-plowing that may be detrimental is bringing zones of calcium carbonate (caliche) or salt accumulation to the surface.

The plowing depth is related to the economic benefits to be gained. Normal plow depth is around 6–8 inches. Some prefer to go to 10–12 inches. It would generally be desirable to utilize as much of the soil as possible in the "plow layer." However, plowing an additional 2 inches may not produce sufficiently increased yield to pay for the extra cost. The depth of plowing may also often be limited by undesirable subsoil characteristics.

"Chiseling," or subsoiling, is practiced in some areas. Its principal purpose is to attempt to break up compacted layers or hardpans that have developed. Chiseling could therefore result in an increased water infiltration rate and possible better water penetration. Another advantage is that it usually leaves the soil surface rough, reducing risk of erosion by wind. Benefits usually last only a year or two. Whether it is desirable to chisel will depend on the need and the economic value. The cost of chiseling has to be evaluated against the value of the extra yield or other benefits that might accrue.

After the land is plowed, it is often "bedded up" with a lister plow (Fig. 3.8). The land is then left in that condition until planting time. The lister plow may also be used for the initial plowing of

3.7. *Deep-plowing is often practiced to improve the physical condition by breaking up a hardpan or other barriers to improve water infiltration, water-holding capacity, and root penetration. (Courtesy Tennessee Corporation)*

3.8. *The land is usually bedded up using a lister plow prior to planting. (Courtesy John Deere)*

stubble ground. After two diskings, the lister will then be used to bed up the land. The lister may also be used to "bust out" the middles of the old rows. Then, within a short period of time, it will be used to "rebed" by plowing out the old rows. The alternative of first breaking out the old row and then rebedding may be practiced. This permits the planting of the row in the same place each year.

The land is often left flat after plowing. It is usually subsequently disked either to control weeds or break up cloddy soils. The land might then be planted. Tillage prior to planting will vary depending on soil type, moisture, and a host of other factors. This is discussed more fully in Chapter 5.

Another practice that is occasionally needed in the irrigated areas following plowing is "land-planing." This is done by a large land-leveling implement designed to take out any minor irregularities that have developed (Fig. 3.9).

Increasing attention is being given to farming systems in which tillage of the soil is reduced to a minimum. Reduced tillage means fewer trips across the field, resulting in lowered costs due to less energy used, less labor, and less wear and tear on equipment. While many producers tilled each field five to eight times in years past, some have reduced that number to three or four trips across a field. In some cases, producers are using a system of no tillage, with the only trip across the field being to plant and possibly apply a herbicide. In a "no-till" system, all of the crop residue is left on the surface and seeds are planted with as little disturbance of the residue and soil as possible. In either case, minimum tillage or no-till, the soil is

3.9. *Land plane used for releveling before land is bedded. (Photo by Leon New, Texas Agricultural Extension Service)*

3.10. *Grain sorghum planted in wheat stubble. (Photo by J. T. Musick, USDA-ARS)*

tilled much less than in most traditional farming operations. In addition to cost savings, minimum tillage also decreases soil erosion and normally increases the water infiltration rate. Minimum tillage also often requires a modification of certain practices, including planting, fertilization, and application of pesticides. Fertilization rates, particularly of nitrogen, may need to be increased, since soil temperatures will tend to be lower due to the surface residues. Surface residues could also harbor insects, resulting in an increased need for insect control. Most of the reduced tillage systems are being adopted by farmers in the higher rainfall areas of the eastern and midwestern parts of the nation. Sometimes grain sorghum is planted in wheat stubble (Fig. 3.10).

Regardless of tillage system employed, the number of tillages is usually kept to a minimum to reduce costs and loss of moisture. Cultivation of the crop after planting has been greatly reduced and even eliminated by some growers by the use of herbicides.

Soil Reaction and Salinity

Soil management includes the maintenance of a desirable soil reaction (pH). It also includes the minimizing of saline-sodic problems in irrigated areas.

Grain sorghum grows satisfactorily over a wide soil pH range. Like other crops, it will produce the best yields within a range of about 6.2–7.8 because some soil factors that affect production are likely to be more nearly optimum within this range. However, grain sorghum has been grown satisfactorily and good yields have been obtained at pH levels from 5.0 to 8.3. When soil pH is below 5.7, fields should be limed to improve nutrient availability and yield. The application of amendments to reduce pH has generally been uneconomical (except for sodic soils); however, certain fertilization practices such as the application of micronutrients may need to be followed. It may be desirable to increase the rate of phosphorus, for example, in order to provide an adequate phosphorus level. A large percentage of the grain sorghum acreage is grown in areas where the pH is neutral to moderately alkaline, in the range of 7.0–8.3.

Since grain sorghum is normally grown in areas where rainfall is low and irrigation is often practiced, salinity problems often develop. This will happen only if the water contains excessive levels of salts or salts that are undesirable.

Two types of problems that may arise from the use of irrigation water are the accumulation of excess salts such as calcium and magnesium in the form of chlorides and sulfates (saline soils) and excess sodium (sodic soils). Excess sodium results in poor physical condition of the soil. Grain sorghum is considered to be moderately tolerant to saline conditions. It is most sensitive to salt at the time of germination and during early growth, but after the 4- to 6-leaf stage, it is much more tolerant.

A soil test is desirable to determine the concentrations of salts and the specific salts present. The proper corrective measures can then be based upon this knowledge.

Leaching with water is necessary to move salts downward if excess salts have accumulated. Where sodium is a problem, either calcium sulfate or sulfur is usually applied. Sulfur is beneficial on sodic soils only if free calcium carbonate is present.

4

Development, Selection, and Production of Hybrid Sorghum

Introduction

Hybrid sorghum, essentially a new crop since the late 1950s, gave a new look to the Western Hemisphere's No. 2 feed grain and the world's fourth most important cereal crop. The discovery of cytoplasmic-genetic male sterility in the early 1950s by Stephens and Holland allowed the plant breeder to modify sorghum's perfect flower into readily manipulated male and female parents. The availability of a totally sterile female and a restorer male, whose pollen was readily wind transported, gave the seed producer a useable system for hybridization (Fig. 4.1). Prior to this sterile system, the crop

4.1. *Close-up of a female panicle branch (left) contrasted with a fertile and male branch (right). The seed from the female produces a fertile hybrid the following generation.*

was mostly self-pollinated, which allowed for relatively easy maintenance of the variety and more trueness to type observed in the field. To the contrary, yield improvement and the accompanying need for various defensive traits were slower and more difficult to achieve with varieties as compared to hybrids. Not to be overlooked and of extreme value to hybrid improvement was the much more extensive entry into the hybrid sorghum seed business by the private sector. With hybrids requiring new seed each year and the necessity of product protection, the seed industry began investing significant sums, supplementing public funds, in research, seed production, and distribution/marketing.

Grain sorghum hybrids are crosses between two unrelated parents but may, for example, with a grazing sorghum hybrid have a single-cross female to improve seed production and therefore be a three-way cross. The grain hybrid is generally of a genetic makeup known as a three-dwarf or waist-high plant height suitable for combining. The grazing type utilizes a sudangrass male parent to produce finer leaved, thinner stemmed, and heavier tillering hybrids grown primarily for grazing, green-chop, hay, or green manure. A second forage type, the sweet and/or silage hybrid, emphasizes high stalk sugar, with utilization being most often under anaerobic conditions of ensiling. All three forms are now successfully hybridized.

HYBRID ACCEPTANCE

In the late 1950s and early 1960s, acceptance of hybrid sorghum approximated an S-curve (Fig. 4.2). Less than 1% of the acreage was

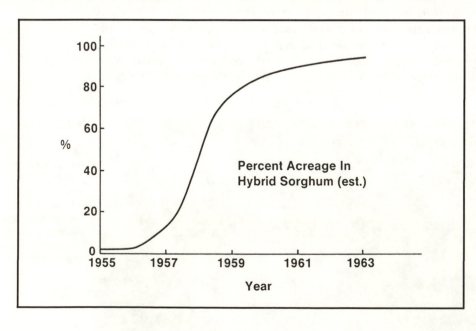

4.2. *Hybrid acceptance by U.S. sorghum producers from 1956 to the mid-1960s.*

planted to hybrids in 1956 but had increased to 10–15% by 1957. Perhaps 75% was in hybrids by 1959, and more than 90% of the U.S. acreage was planted to hybrids by the early 1960s. Essentially all the grain sorghum acres are now hybrid, while the forages still have a limited acreage in varieties that have been grown for a good many years. Contrast this with corn, where only 17% of the corn in Ohio was hybrid by 1942, a time lapse after hybrid introduction equivalent to 95% sorghum acceptance. No doubt the success of hybrid corn greatly helped remove any skepticism the producer may have had for hybrid sorghum.

YIELD TRENDS

During the 30 years of hybrid sorghum (1956–1985), sorghum yields have increased from 22.1 bushels per acre to the record high 66.7 bushels per acre in 1985. This range followed a 1955 prehybrid yield average of 18.9 bushels for an average annual gain of 1.59 bushels a year. This is 3.2% of the 30-year mean yield of 49.5 bushels per acre. The correlation of yield to years during this early history of hybridization was $R = 0.76$ for a highly significant relation. Approximately 58% of this yield improvement may be attributed to the cultural and genetic changes that have occurred over this time frame.

Table 4.1 summarizes U.S. yields in 7-year increments, a period considered by many to be the average hybrid life span. Note that the first two increments covering the 1944–1957 period essentially reflect varietal yields, whereas the 1958–1985 period is basically all hybrid performance. No doubt the availability of a better product and at a higher cost encouraged a much more serious approach to the crop by the producer. If we attribute heterosis (hybrid vigor) to account for one-third of the early gain, or perhaps 7 bushels per acre, then fertilizer, herbicides, irrigation, and other factors would account for 13–14 bushels per acre, and this combined 20-bushel increase pretty well explains the doubling of yields during the 1958–1964 period compared to varietal averages prior to that time.

The next 7-year period (1965–1971) reflects a time when new germplasm and hybrid breeding methods replaced early hybrids,

Table 4.1. U.S. sorghum yields in 7-year increments, 1944–1985

Years	Acreage yield
	(bu/acre)
1944–1950	18.7
1951–1957	20.4
1958–1964	41.1
1965–1971	52.7
1972–1978	53.4
1979–1985	57.5

which were all too dependent on available varieties. This time frame saw improvements such as increased drought tolerance, improved stalk quality, wider maturity range, improved grain quality, yield/maturity class, and virus tolerance/smut and anthracnose resistance.

The 1972–1978 period shows little gain in yield, with a reduced release of new hybrids as the industry hurried to develop greenbug resistance as well as attempted to find adequate combinations of downy mildew resistance and stronger performance for hybrids in the South, especially Texas. This same time frame saw significant shifts of better and often irrigated land away from sorghum and toward corn, resulting in reduced inputs to the crop where it was grown under less favorable conditions.

Finally, the most recent period (1979–1985) averaged 57.5 bushels per acre, or nearly three times more than the prehybrid 20-bushel and less yields. Interestingly, this recent increase came at a time when producers faced a difficult financial dilemma and frequently had attempted cutbacks in production inputs. The trend over the 30 years can best be illustrated by Figure 4.3, which points out the growing importance of genetic gain in relation to new cultural inputs. Certainly, hybrids with more drought and greenbug tolerance as well as more tropically adapted germplasm for the South have made a positive impact on sorghum performance in recent years.

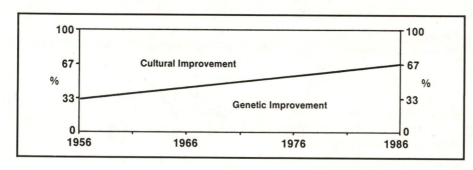

4.3. *Changing ratio of sorghum improvement from 1956 to 1986, with increasing emphasis on genetic gain.*

Hybrid Development

As a mostly self-pollinated crop, having a perfect flower (male and female in same flower), sorghum evolved for thousands of years, with each crop resembling the previous generation. Changes only occurred by a low frequency of chance outcrossing and the infrequent mutation that might have a desirable advantage, allowing it to remain in the population. Selection and breeding pressure in the prehybrid years emphasized earlier and shorter plants that could adapt to the more temperate Great Plains and exhibit improved standability and drought tolerance while yielding better than the commonly grown cultivars. Where sorghum was confined to the drier plains states, disease problems were minimal, although milo disease was selected against and charcoal rot remained more or less uncontrolled. Chinch bugs at times caused serious economic loss to the crop, while midge damage was confined to more southerly sorghum-producing areas. Much of the basic germplasm affecting varietal development (some 30 sorghums) came into this country from 1880 to 1910. The yellow endosperm types, however, from Nigeria and the Sudan did not become a part of breeding programs until the early 1950s.

At this same time, Stephens, who had been looking for a usable sterile system since the 1930s, working with Holland, discovered the cytoplasmic-genetic interaction that allowed for the production of 100% male-sterile plants (Fig. 4.4). This unique interaction of a

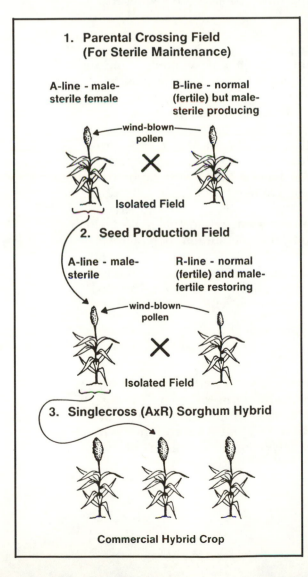

4.4. *Commercial hybrids are produced by (1) first producing a male-sterile female line, which is then (2) crossed with a line that is normal and that restores the sterile female to produce (3) the fertile single-cross commercial hybrid.*

1. Parental Crossing Field (For Sterile Maintenance)

A-line - male-sterile female

B-line - normal (fertile) but male-sterile producing

wind-blown pollen

X

Isolated Field

2. Seed Production Field

A-line - male-sterile

R-line - normal (fertile) and male-fertile restoring

wind-blown pollen

X

Isolated Field

3. Singlecross (AxR) Sorghum Hybrid

Commercial Hybrid Crop

4.5. *A typical seed field on the Great Plains grown under low humidity and irrigation and with sufficient isolation for good seed purity.*

sterile cytoplasm (cell sap outside the nucleus) with the genetic code for male sterility (genes in the nucleus) resulted in a nonpollen-producing plant that could serve as the female. When crossed by a fertile and restoring male type, the resulting hybrid was fertile and, depending on the diversity of the parents, could be considerably higher yielding than the best parent.

The cytoplasmic aspect is essential since only the nuclear genes, not the cytoplasm, move through the pollen. Thus the nonrestoring parent (B line) is fertile from its normal cytoplasm, whereas it only contributes sterile genes through the pollen. The female contributes all of the cytoplasm (which is sterile) plus sterile genes and thus continues to produce 100% sterile offspring or female parents.

With wind moving the relatively light sorghum pollen, a common sight in seed-producing areas (Fig. 4.5), beginning in the mid-1950s, is 4 rows of pollen-shedding male and 12 rows of male-sterile female. Other combinations may be 6:18 and 8:24, with ratios depending on row width as well as on pollen shed and splits between parents. Only grain from the female rows will be hybrid and saved for conditioning as commercial planting seed, while the male becomes commercial grain. Unfortunately, this phenomenon explained above, which changed sorghum from essentially a self-pollinated crop to a cross-pollinated crop such as corn, makes the end product vulnerable to wind-blown pollen from sorghums other than the desired male. These weedy or off-type sorghums will be discussed in detail in Chapter 5.

BREEDING PLATEAUS

The sorghum breeder attempts to relate available germplasm (genetic source material) to the specific environment where the hybrid will be grown. Progress may be evaluated in terms of yield or defensive plateaus, with either or both associated with an improved product. Following is a generalized analysis of these periods of improvement:

1. 1956–1964. Most of the early hybrids marketed—there were some notable exceptions—relied on male-sterile and restorer parents that had been developed for and were also utilized as varieties. They exhibited adequate hybrid vigor to provide the incentive for much improved cultural practices (Fig. 4.6). U.S. yields doubled over pre-hybrid years. Their weaknesses were numerous, however, in that stalk quality, disease and insect resistance, and adaptation were mediocre at best. Also, their genetic base was narrow and thus vulnerable as a group to changes in the environment.

4.6. *The variety, Westland, growing in western Kansas (left) compared to a commercial hybrid (right), with the hybrid being 4 days earlier and 48 percent higher yielding.*

2. 1965–1971. This period saw an additional 28% yield increase and reflects the first hybrids having parents developed specifically to exhibit hybrid vigor. Additionally, the breeders made rather significant improvements in the nutritional value of sorghums, with much credit due Webster and Karper. These newly introgressed African-origin introductions with yellow endosperm were developed into adapted, more temperate germplasm (Fig. 4.7). Although numerous additional traits, e.g., drought tolerance, were associated with this breeding improvement, the significance of the effect of yellow germplasm is often overlooked on lengthening the postbloom or fill period, resulting in a significantly improved test weight.

3. 1972–1978. As noted in Table 4.1, yield was minimally enhanced during this period of regrouping by breeders. This 7-year increment may, however, have been the most significant breeding effort to date, as germplasm from world collections (which number

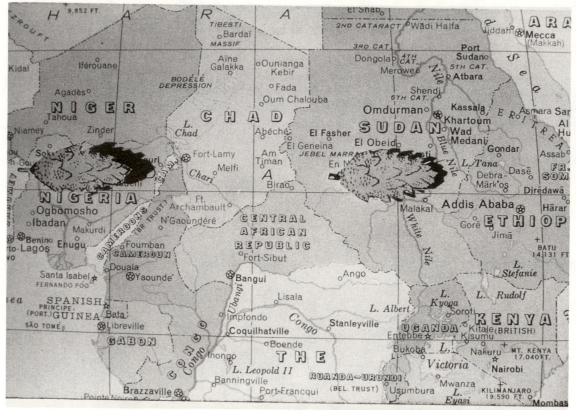

4.7. *Plant introductions from Nigeria and the Sudan became a significant source of yellow endosperm breeding material after introgressing them into short, photoperiod-insensitive types.*

more than 25,000 varieties) was rapidly screened for disease and insect resistance. In the process, many additional favorable traits were located. Ethiopian sorghums provided a much needed resistance to downy mildew, with essentially all hybrids grown in problem areas now carrying adequate levels of resistance to the first pathotype, and breeding is under way to incorporate resistance to the newer or pathotype 3 strain.

In the process of incorporating downy mildew resistance, much was learned as to the need for tropical adaptation to improve yields where the crop is grown under shorter days, at low elevations, and with high humidity and warm nights. Sorghums grown under such conditions are now much better adapted to the environment, making them of higher and more consistent performance.

The greenbug change to biotype C and from small grain to sorghum in 1968 was a real threat to the acceptance of the crop and no doubt explains in part a growing use of corn on acres previously used for sorghum (Fig. 4.8). Public and private research, using the world collection and especially a tunisgrass sorghum, had enough seed with resistance to the insect to plant 4 million acres by 1976. Due to a diverse genetic pool used to control biotype C, fortunately only half or less time was needed to convert to biotype E resistance.

4.8. *Sorghum susceptible to biotype C of the greenbug (foreground) and a resistant hybrid (background) showing normal plant development.*

4. 1979–1985. This last plateau reflects the renewed effort for yield improvement, being 8% higher than the previous level and some three times higher yielding than the prehybrid plateaus. The breeder now concentrates on resistance to new forms of greenbugs, anthracnose, smut, and mildew plus continued efforts to develop midge resistance in hybrids without sacrificing yield potential. In addition, the demand for more drought tolerance grows greater each year as does the interest in nutritional improvement. Also, we are seeing much improved hybrid performance in the fringe or marginal areas, such as by very early maturing hybrids for the North with good standability and feed value. Even so, the challenges remain much greater than the accomplishments.

BREEDING FOR YIELD

It was pointed out in Figure 4.2 that as producer inputs remain constant or even decrease, yield improvements must depend on gains from genetic inputs through breeding. The conversion to hybrids, beginning in the mid-1950s by allowing the producer the benefits of heterosis, represents the most significant breeding accomplishment for yield. Shull in 1913 coined the term heterosis as a way of expressing the F_1 hybrid exceeding the parents with respect to some characteristic such as size or rate of growth. Heterosis may frequently be described as hybrid vigor. Yield results from the input of many genetic factors or genes and is thus termed "quantitative" as opposed to a simply controlled trait that would be qualitative in its inheritance, such as greenbug resistance. No one knows, but it is speculated that as many as 5000 genes may affect yield. In addition, the best efforts of the breeder may be masked if the necessary defensive traits are lacking. The complexity of yield may be further illustrated by the fact that only 20 genes can be arranged in some 3.5 billion different genotypes or genetic combinations that would occupy more than 50,000 acres.

To breed for yield means increasing the diversity between parents and in turn maximizing the favorable-yield genes. Sorghum breeders rely heavily on a world collection that exceeds 25,000 varieties. Various breeding techniques also allow the breeder to rearrange genes already available. Finally, a third approach would be to increase the level of defensive traits, allowing for the increased expression of yield genes (Fig. 4.9). As research programs increase in quantity and quality, the trend for increasing yield should continue. The producer should also recognize variations in yield stability. Some hybrids excel at low-yield levels with stress but are limited at

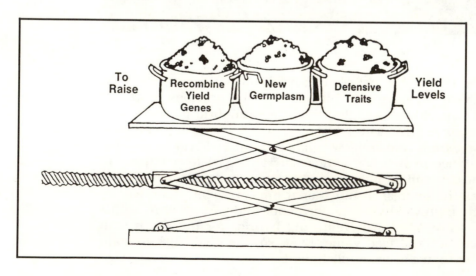

4.9. *Breeders strive to move sorghums to higher yield plateaus by recombining available genetic material for more heterosis, introducing new germplasm from the world collection, and breeding in as many defensive traits as possible.*

optimum levels. Others exhibit a reverse reaction. Finally, some that are quite stable perform relatively well under both stress and optimum environments.

NUTRITIONAL IMPROVEMENT

The first sorghums grown in the United States were basically red or white seeded, although the hegari type, for example, was characterized by a subcoat containing tannins (phenolic compounds), as were the brown sorghums. Tannins are known to depress protein and starch digestion but occur at minimal levels in most sorghums or are essentially absent in the yellow endosperm types. Historically, sorghums were considered in the prehybrid and early hybrid years to be only 88% the value of yellow corn. In the 1960s, however, as yellow sorghums were more commonly incorporated into the newer hybrids and with better grain-processing techniques, daily gain and feed efficiency values improved dramatically. The National Research Council, in their *Nutrient Requirements of Beef Cattle*, rates the net energy value of flaked grain sorghum at 96–97% of flaked corn. Numerous studies would suggest that little difference now exists between the relative feed values of corn and sorghum, depending somewhat on the class of sorghum.

In areas of the world where sorghum is primarily a food rather than a feed grain, more concern has been placed on grain type. The U.S. breeder has also been upgrading the quality aspects of the hybrids and almost entirely phasing out the brown types. Additional opportunities exist with high-lysine and waxy sorghums plus the diversity of nutritional quality from the large world collection. The advent of red × yellow and yellow × yellow sorghums has given additional support to the improved nutritional levels of sorghum grain. The yellow color (Fig. 4.10) is also important to poultry producers because it affects yolk appearance and skin color. Studies

4.10. *In recent years increased acreages of cream and yellow hybrids are obvious at county elevators.*

additionally support improved protein quality such as glycine, lysine, methionine, and valine with the true yellow types. Both digestibility and palatability seem also to be improved in certain yellow endosperm types. The newer hybrids have an improved ratio of corneous to floury endosperm, which is reflected in higher test weights.

STANDABILITY

This defensive trait, along with early maturity, at times becomes as significant to the producer as yield potential. To simply increase yield without regard to an improved stalk makes the hybrid more vulnerable to lodging. By using stiff or shorter stalked breeding material and incorporating more drought tolerance (especially of the postbloom type) with as much disease and insect resistance as possible, breeders have made dramatic improvements in standability. This breeding objective will remain of high priority in the future.

MATURITY

With control over a valuable system of maturity genes, the breeder is able to provide a "quick hybrid" for the short-season area, for drought conditions, or for double cropping (Fig. 4.11). The medium-early to medium maturity class, by far the most popular, seems most likely over the years to produce a consistently strong crop in areas faced with some seasonal stress. The medium-late to late class requires more water and fertility. Also, the later hybrids may be subjected to heavier midge pressure in the South. With all factors optimum, this maturity category gives the opportunity for maximum yield. The breeder, however, develops hybrids with increased yield potential within each maturity class rather than depending on maturity as a source of yield.

4.11. *Earlier maturing sorghums have moved sorghum hybrids farther north and made them more suitable for double cropping with wheat. This July 5 planting in south central Nebraska has reached maximum dry weight in only 77 days.*

DROUGHT AND HEAT TOLERANCE

Sorghum is essentially a dryland crop grown in an area of less favorable moisture and often higher temperatures than the primary corn-growing areas of the United States. For consistent or stable production, a hybrid needs all the stress tolerance possible without sacrificing yield potential when conditions are favorable. The breeder in conjunction with the physiologist can utilize several mechanisms of drought and heat tolerance (Fig. 4.12):

Primary	*Secondary*
1. Leaf water loss	1. Maturity
2. Postbloom tolerance	2. Charcoal rot tolerance
3. Root development	3. Greenbug resistance
4. Heat tolerance	4. Nontillering habit
5. Dormancy	5. Grain:stover ratio
6. Osmoregulation	6. Heterosis

However, the complexity of drought tolerance and its interaction with a variable environment make this defensive trait important and yet a difficult challenge to the breeder. Currently, the breeder's best progress against drought involves working diverse germplasm from the world collection, followed by a comprehensive selection and testing program against a wide range of stresses. The introduction of hybrids having higher yield potential illustrates a form of improved water use efficiency.

4.12. *Drought- and heat-tolerance improvements have been incorporated into modern hybrids and better understood through much input by plant physiologists. Here, Dr. Charles Sullivan, University of Nebraska, utilizes a porometer to measure leaf water loss.*

DISEASE RESISTANCE

Most commercial hybrids carry tolerance or resistance to one or more diseases. While the first 10 years of hybrids saw minimum disease problems aside from stalk rot and head smut, by the late 1960s and early 1970s, the sudden appearance of sorghum downy mildew stimulated a whole new breeding effort (Fig. 4.13). Now this disease has been further confounded by another virulent pathotype.

4.13. *Downy mildew of the systemic form first appears as a chlorotic striping of the leaves followed by a complete disintegration of leaf material and an absence of grain production.*

Fortunately, some Ethiopian sorghums contain a good and simply inherited source of mildew resistance. First, however, breeders conducted a conversion program to make this source material shorter and photoperiod insensitive for temperate regions. Now, resistance to mildew has become a basic requirement in both grain and forage hybrids, thus changing the whole hybrid picture in the South during the 1970s. Besides downy mildew, sorghums grown in the more humid areas are bred for numerous other foliar diseases such as anthracnose, rust, and zonate leaf spot. Anthracnose, like mildew, continues to have changing pathotypes, which makes breeding more difficult.

Stalk rots such as charcoal rot and *Fusarium* relate rather closely to moisture and temperature stress (Fig. 4.14). Drought tolerance is critical to keep the plant healthy and photosynthesis and

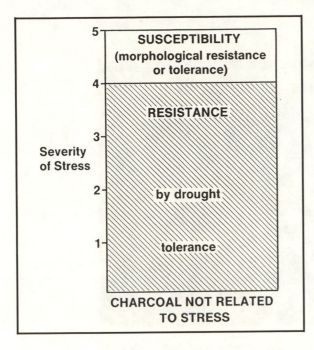

4.14. *Stalk rots relate closely to postbloom stress tolerance, with considerable variability between hybrids. Sorghums with a stay-green characteristic show less tendency for these losses.*

translocation functioning normally. When plants stress, senescing root cells are invaded by microorganisms until water uptake is reduced and permanent wilting occurs. Stalk rot organisms can now destroy the remaining stalk structure, which results in lodging. By inoculating new parents and hybrids with these organisms, the breeder can screen for resistance. Perhaps the primary gain in handling stalk rots, however, has been to incorporate more drought tolerance, which in turn allows for normal plant functions.

Virus tolerance or resistance has been a significant breeding objective since the mid-1960s. Hybrids should show no more than leaf mottling to avoid serious loss from this disease. The intermediate hosts, such as johnsongrass, serve as a source of inoculum for virus as well as most sorghum diseases.

INSECT RESISTANCE

No doubt the availability of intermediate hosts such as johnsongrass and *Sorghum almum* also intensifies problems with the most serious insect pest of the South, the sorghum midge. In the 1970s, however, germplasm from Ethiopia and Brazil showed considerable promise of limiting seed loss and reducing the reproductive capacity of the insect. The mechanism, although not fully understood, makes the deposit of eggs more difficult. Additionally, nonpreference and antibiosis are present, the latter term referring to a reduction in the reproductive capacity of the insect. The GS-3, or grain-fill period,

4.15. *Midge resistance (right) offers the producer a defense mechanism and reduces management input as well as production costs.*

moves more rapidly with midge-resistant hybrids, allowing the ovary to develop quickly and perhaps in this manner reducing larval damage.

The level of midge resistance varies considerably and seems most effective when both parents are resistant (Fig. 4.15). Threshold requirements for spraying now appear to be 5–10 times greater than with normal hybrids, which should be sprayed when a single midge per head is observed. The breeder, while striving for higher levels of resistance, cannot ignore yield and other desirable traits, since the presence of the insect is uncertain. To the contrary, extensive testing has shown the dramatic yield increase possible with resistant material in the absence of chemical control (Table 4.2).

Table 4.2. Midge trials at Salto, Prov. de Buenos Aires, 1981

	Yield	
	Test 1	Test 2
	(lb/acre)	
Resistant hybrids (24)	4794	3945
Susceptible hybrids (11)	2042	1729
Percent yield loss	57	56

The greenbug of cereal crops mutated to biotype C and did serious damage to sorghum beginning in 1968 (Fig. 4.16). Fortunately, a dominant gene for resistance from tunisgrass was readily incorporated into parent lines, greatly reducing the damage from this sucking insect that causes necrosis of leaf tissue by injecting a toxin into the plant. Besides this tolerance mechanism, the resistance gene in hybrids also displays nonpreference and antibiosis.

A significant supply of hybrid seed, enough to plant 4 million acres, was available for the 1976 crop. However, in 1980, a new biotype called "E" appeared, causing a renewed effort by breeders to use other world collection sources of resistance. E-resistant hybrids were available by 1983, much more quickly than with the C biotype. While genetic resistance has numerous benefits, ranging from improved yield to better leaf digestibility, part of the attention of the breeder, unfortunately, is diverted from yield breeding to keeping ahead of a changing biotype.

4.16. *Since 1968 considerable effort has been given to incorporating resistance to the sorghum greenbug, with thousands of seedling rows screened annually. Larry Lambright of DeKalb Pfizer Genetics inoculates a greenhouse flat and will read the material in approximately 14 days.*

ADAPTATION AND TESTING

As with all crops, sorghum responds closely to its environment, which is unfortunately not consistent even at a specific location. For example, daylength, moisture, humidity, heat units, and the disease and insect complex facing a given field will vary by planting date, weather pattern, and the presence of intermediate hosts. In addition, soil fertility, soil pH, soil type and structure, elevation, plant population, row width, and temperature extremes plus many other factors play an important role in hybrid performance.

An example of adaptation receiving considerable attention and one that is applicable to the Western Hemisphere has been termed "tropical adaptation." A hybrid of this category will have a significant amount of its genetic component from converted tropical sources of breeding material that allow for photoperiod insensitivity, tolerance to higher humidity as well as night temperatures, improved performance at low elevations, and considerable disease resistance. These traits are helping sorghums show much improved performance in the southern United States as well as in Central America, Mexico, and much of South America, which is tropical or subtropical. The first hybrids, to the contrary, were of a heavy temperate adaptation, with yields adversely affected by warm nights, shorter days, and high humidity, the latter condition leading to more disease problems.

Average yields of sorghum are perhaps one-third or less of record yields. While insects and disease may account for about 15% of the difference, the remaining 85% is related to environmental stresses that mask the available genetic potential for high yield. The breeder faces a high challenge of meshing germplasm with the environment, including cultural practices. The best solution is maximum testing of experimental hybrids over diverse environments before commercial release.

Optimum testing evaluates material over as many environments affecting the principal producing areas as possible (Fig. 4.17). Hybrids should be screened under the current or expected cultural practices. To overcome the effect of years on performance, a minimum of 4 years of research data helps insure the release of consistent products. Tests are replicated, meaning each hybrid appears more than once to help take out variation within the trial.

Because of the possible permutations within the genetic system of sorghum and the seemingly infinite number of potential parental lines, sorghum breeding truly becomes a numbers game. For example, 400 new greenbug-resistant females and 800 resistant males theoretically could generate 320,000 new hybrids for evaluation, a nearly impossible chore. To the contrary, a well-designed evaluation program, which effectively sorts out the most desirable pedigrees, has the same importance as the developmental phase already discussed.

4.17.
Thousands of new sorghums are tested annually over many locations with only a few ever reaching the producer.

A range of moisture and soil conditions across locations allows the breeder to determine stability of performance, which greatly affects the selection of a new hybrid. By comparing yields of an entry to the means of the tests in which it is evaluated, an indication, or "b value," can be determined that reflects the response of the pedigree to varying yield levels. As shown in Figure 4.18, some hybrids will

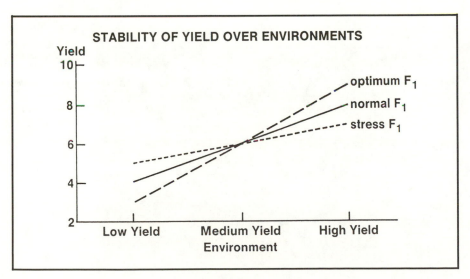

4.18. *Yield stability will give the producer consistent production whether conditions be average, optimum, or stressful.*

outyield others under less favorable conditions but will have yield ceilings somewhat below the maximum expressed by another pedigree under more favorable situations. In the uncertain environment faced by most sorghum producers, an intermediate category (slope b = 1) will offer the best opportunity for a profitable crop.

While the above discussion attempts to include principal breeding objectives, many others are significant. Certainly, in some areas grain weathering resistance has high priority. Seedling vigor is often critical to stand development. Selection of parental lines tolerant to more acid soils has recently become of higher priority as the crop moves east. Other traits such as head exsertion, plant height, and ease of seed production are routine considerations for the breeder.

BENEFITS OF INTERNATIONAL RESEARCH

To meet the objectives previously discussed, today's plant breeder constantly attempts to be more efficient and to gain an extra generation or two whenever possible. All breeding programs have winter nurseries, which may be at low-latitude locations or in the Southern Hemisphere. Here, much of the breeding for another generation can take place; in fact, a shorter daylength or different disease complex may even expedite the breeding effort. Certainly, to work the tropical photoperiod-sensitive lines, these winter nurseries are a great asset. Testing and release of new, improved hybrids have no doubt taken place 2–4 years earlier than normal because of a second crop each season.

Sorghum efforts in other countries often occur during U.S. winters, thus giving a second generation of observation, crossing, or line increase (Fig. 4.19). These foreign programs allow for screening for

4.19. *Winter sorghum nurseries (this one in Puerto Rico) greatly speed up the development of new and improved hybrids.*

insect resistance such as midge and greenbug. Certainly, an experimental hybrid of promise in the United States that shows poor resistance to a disease in Argentina should be evaluated more intensively prior to release (or possibly not released to all), such as with downy mildew breeding. International breeding projects often lead to a more ready exchange of germplasm or techniques that can benefit the producer back home. A significant benefit of a reciprocal nature relates to foundation seed increase, which often can be done in quantity only under more temperate conditions in either hemisphere. A winter increase in Argentina can put parent lines into U.S. production fields only a month or so after harvest. Sorghum breeding for the U.S. producer is no longer a once-a-year activity; in fact, a crop is growing every day of the year.

Selecting a Sorghum Hybrid

Radio, television, newspapers, magazines, mailings, test plots, and discussions with other producers make it obvious to the grower that there is a large selection of available hybrids. In fact, the offering is so large, the producer may be confused as to which hybrid, or hybrids, to choose. Although personal preferences as to brand and type of hybrid affect the buying decision, a few suggestions follow:

1. Choose a hybrid with stability of performance over years and environments. A consistent performer with maximum defensive traits such as drought tolerance will be more successful than a 1-year wonder. Include standability as well as yield in your decision making.

2. Where inputs including irrigation and fertility are optimum, look for a hybrid with potential for high yield in its maturity class.

3. Attend field days and off-season product meetings. Ask questions.

4. Choose a dealer in which you have confidence, one who is supported by a company with hybrid and cultural information.

5. Look closely for a hybrid from a company you respect as being research oriented. This company likely will have the most improved product available more quickly.

6. Look at your neighbors' crops, especially those of sorghum growers you respect as outstanding. Chances are they face environments most similar to yours.

7. Study plots and data generated by public and private research. Preferably look at data generated over years and results from more than one location. Consider the maturity you desire as well as specific disease and insect resistance.

8. Consider the purity of the seed and also the germination. Ask the dealer about the number of seed per pound and adjust planting accordingly.

9. Try to eliminate price in your decision making. Seed is one of your least expensive inputs. The range in cost per acre between hybrids will be less than the value of a hundred weight of the end product. Seed costs average only 3–4% of the gross return.

10. If you feed your crop, give consideration to the nutritional value of the grain.

11. Keep records throughout the season, including specific hybrid planted (not just brand), performance throughout the season, and relative yield related to your other choices.

12. When planting a new hybrid, avoid changing all your acres at once. Also ask the dealer about any special characteristics that will enable you to better handle the new pedigree and avoid surprises during the season.

13. Know if the hybrid has a new and particularly helpful trait such as disease and insect resistance; the performance in the absence of the pest should be competitive to your previous favorite.

Yield data should be accompanied with appropriate statistical analysis to indicate variation and validity. The coefficient of variation (CV) expresses the variability in a trial due to error in relation to the mean or average yield of the test. A CV above 10 with irrigation or above 14 with dryland data suggests excessive variability that could make results misleading. Under stress with a low mean yield and more plot variation, CVs will be higher.

There are various ways to show significant differences between hybrids that may require a knowledge of statistics to properly interpret. Frequently, a P value of 0.05 is reported with test results where significant differences are indicated. This terminology simply refers to the probability that the given significant difference (e.g., 800 pounds per acre) would occur no more than 5% of the time due to chance alone based on the variability within the trial.

Often a hybrid selection will be preferred that is of an earlier maturity than the highest yielder in a test. These hybrids may show more consistency over years by requiring less water and other cultural inputs, plus they may escape an insect or early frost problem or even gain a better market price. To consider these earlier test entries, the yield per day from planting to flower may be the easiest calculation on which to base a decision. Remember that hybrids with improved offensive and defensive traits will continue to be developed in a competitive seed industry. After careful evaluation of available data and field observations when possible, try a few bags of a newer hybrid for comparison with your current favorite sorghum.

Production of Hybrid Sorghum Seed

To produce hybrid sorghum seed requires (1) pure breeding stocks of parental seed, (2) adequate wind for pollination, (3) identical blooming of both male and female, (4) adequate isolation, and (5) harvest conditions allowing for grain of high test weight, germination, and vigor.

FOUNDATION SEED

Pure breeding seed stocks depend on a research program that is competent in providing a perfect source of breeder's seed to the foundation program. This may only be enough seed for 1 acre, which is adequate for an initial increase, since 6 pounds can readily be increased to over 4000 pounds. Even with maximum attention, a breeding stock needs frequent new starts and subsequent cleaning up to maintain sterility, uniformity, and type. Some sorghums have a very high mutation rate for the third height gene, which requires new increases from research to reduce the frequency of taller plants in the commercial crops. Also, foundation increases require even better isolation from other sorghums and more equipment cleanliness than with commercial seed production. With smaller blocks, more hand rogueing as well as hand harvesting can take place.

COMMERCIAL SEED

Seed production in the United States takes place on 40,000 or more female acres, primarily in regions of expected good seed yields, minimum weathering, and reasonable isolation. The ratio of female rows to male rows varies with pollen-shedding capacity and tillering of the male as well as wind conditions typical of the area. A ratio of 3:1 is common. Theoretically, each male plant produces approximately 45 million pollen grains, which would be adequate for 15,000 female plants; however, most miss the target, thus requiring a dense level of pollen on the critical days of flowering.

Often the parents of the best performing hybrids do not "nick" or bloom together, which means planting the seed field at two or more dates. This nick problem adds to the uncertainty of good seed set and more chance for contamination, makes the production more costly, and frequently suggests two planting dates for the male to better insure good pollination.

Seed production free of high humidity or continuous rains at maturity gives the highest quality seed. Quality is almost always better with the harder seed or seed with a high tannin or brown subcoat. Fungicides may be significant in upgrading the quality of foundation or commercial seed production.

At the time of contracting seed acreage with growers, good cultural practices, and most important, fields isolated from sources of contamination, are considered. Rogueing fertile plants in the male-sterile female as well as off types in either parent requires a dedicated and large labor force. Surprisingly, much of the rogueing cost often goes for cleaning up outside the actual seed field, such as in roadsides and nearby cornfields. Regular inspection during bloom and a final rogueing just ahead of combining provide information on the success of the nick, purity of the parents, and a last chance to remove weedy sorghum that may have come back late in the season. Seed companies set high standards of isolation to minimize the frequency of off-type or weedy sorghums.

Immediately after harvest, seed producers send combine samples to the South, often to more tropical countries, for winter growouts. As soon as possible, the various production lots are thoroughly studied for purity, with standards for disqualification, including germination, making a final determination on the amount of available planting seed. The seed are cleaned, treated with a fungicide, safened where appropriate, and placed in protective bags. The various seed-conditioning centers then transport this seed to the many seed dealers (Fig. 4.20).

The hybrid sorghum industry in the United States, from development to intensive testing (allowing for hybrid selection) and finally quality foundation and commercial seed production, exemplifies progressive agriculture's effect on gains in yield per acre and total production.

4.20. *Following conditioning, seed are placed in protective bags and shipped to dealers and producers.*

5

Cultural Practices

THIS CHAPTER DEALS with the actual production of grain sorghum. Many of the important cultural practices will be discussed either completely or in part elsewhere in the book. Because specific cultural practices vary greatly from place to place, only important principles and general practices will be discussed here. It is the proper combination of practices that produces profitable yields.

Variety of Hybrid and Seed Choice

The modern sorghum hybrids have virtually replaced varieties. Hybrid vigor and the opportunity to incorporate desired agronomic characteristics into a hybrid have been responsible for this change. The large number of hybrids available makes selection difficult for a particular locality, soil, or farm (Fig. 5.1). No single hybrid will possess all the desirable features that a grower or processor wants.

5.1. *The proper choice of quality seed of an adapted hybrid results in a uniform crop and high yields.*

63

Considerations in choosing a hybrid are discussed in Chapter 4. The variety or hybrid chosen should fit the growing condition. If moisture is adequate and yield potential is high, the hybrid must also have a high-yield potential. If moisture is limiting and yield potential is low, choose a hybrid adapted to that situation. Many of these factors are interrelated. For example, lodging is often caused by the disease called charcoal rot, which is most prevalent during or following drought stress. In erratic rainfall areas, the risk of severe moisture stress increases as the length of the growing season increases. Therefore, the probability of charcoal rot is greater when longer season hybrids are grown. In this case, quick-maturing hybrids may give the highest harvestable yield even though the yield potential of longer maturing hybrids is larger.

The selection of the proper hybrid in fringe production areas is more important than in the intensive production regions (Fig. 5.2). In the more humid areas it is important that hybrids dry rapidly. Fast-growing hybrids with loose, open panicles usually selected for humid regions will normally have less insect activity (Fig. 5.3).

5.2. *Modern grain sorghum hybrids grown in the major semi-arid and arid sorghum belts have long, tight heads.*

Factors important for seed selection are germination, presence of weed seed, hybrid purity, crop mixtures, and appearance. If any given company does not adhere to good seed standards, perhaps other hybrids with similar characteristics should be chosen.

Seed treatment is now widely used. Most sorghum is treated with a fungicide. Good-quality seed are seed high in food reserves. In general, seed high in protein give higher germination and seedling vigor than seed low in protein. It has been concluded from research studies that better germination (higher percentage, more uniform emergence) and sturdier plants result from the use of plump, high-quality, high-protein sorghum seed.

Seedbed Preparation and Planting

The objective of a good seedbed is to provide an optimum environment for germination and seedling establishment. An ideal seedbed is one in which the seed can be placed into firm, moist soil and covered with soil that is moist but loose. The seedbed should be free of weeds and tilled in order to deter weed germination.

It is not necessary that the entire field be prepared for optimum seed germination. Only the part of the row where seed are to be placed needs to be conducive for germination. In fact, the rest of the area should be loose so as to deter weed germination. Therefore, two zones need to be considered. One zone, known as the seed zone, needs to be firm; the second zone, the area between the rows known as the water management zone, needs to be loose to enhance water infiltration and deter weed seed germination. In addition, the entire soil area below the surface serves as the rootbed. It is not reasonable

5.3. *Sorghums with looser, more open heads are usually selected in humid areas.*

to prepare the entire area for germination when such a small percentage of the area is used for that purpose.

There are several systems commonly used in seedbed preparation. Sorghums are commonly planted on beds under irrigation and in the furrow or flat under dryland conditions (Fig. 5.4). There are several bed types in use for irrigated sorghum. These include variations from one row on top of the bed on standard 38- to 40-inch spacings to three rows on the bed. Bed tops must be wide enough so that the rows are completely on top of beds. Lodging is often encountered when the rows are at the sides of the bed. Planting two rows spaced 17 inches apart on top of a 54-inch bed is used in some areas (Fig. 5.5). This gives an average row width of 27 inches. This system allows for wider irrigation furrows, which facilitate watering, especially on long runs.

In dryland sorghum production, the seed are sometimes planted with a grain drill on a flat seedbed. Row widths vary from 10 to 30 inches depending upon row spacings possible with the grain drill. Conventional grain drills used in most sorghum-growing areas are 6, 8, 10, 12, or 14 inches. Any multiples of these spacings can be used. The most common ones are from 14 to 24 inches. When drilling sorghums, a slightly higher seeding rate can be used in comparison to conventional wide-row spacings.

There is a trend toward drilling sorghums when sprinkler irrigation systems are employed as compared to multirows on the bed for furrow irrigation.

The narrower row spacings give higher yields than the old standard 38- to 40-inch rows. Most experiments indicate that a row spacing of 20 inches is about optimum. In many irrigated areas there

5.4. *Under dryland conditions, grain sorghum is usually planted in furrows.*

5.5. *Irrigated sorghums are usually planted with multiple rows on beds.*

is a compromise between optimum row spacing and spacings that facilitate good row or furrow irrigation.

Broadcast sorghums, where each plant is equally spaced, is theoretically the most efficient seeding method. Broadcast sorghums, however, are difficult to irrigate when furrow methods are used. Dryland as well as irrigated sorghum tends to sucker more and lodge more freely when broadcasted or drilled. As hybrids are developed that reduce these problems, drilling and broadcasting may increase.

RATE OF PLANTING

The rate of planting depends upon the yield potential, which is primarily controlled by available soil moisture. In general, it requires about 10 heads per pound of seed yield expected at harvest. Therefore, a population of 1000 plants per 100 pounds of potential yield is needed. As seen in Table 5.1, as the plant population is increased, row spacings must decrease. Plants too thick in the row will produce smaller than normal heads. Whenever yield potentials exceed 3600 pounds per acre, row spacings less than 40 inches should be used. Whenever very high yield levels are being sought (above 7000 pounds per acre), row spacings averaging less than 30 inches are recommended.

Table 5.1. Estimated seeding rate and recommended row widths based upon yield potential

Yield potential	Plant population	Row spacing	Seeding rate
(lb/acre)	(plants/acre)	(in.)	(lb/acre)
1800	18,000	40	2.4
2600	26,000	40	3.4
3300	33,000	40	4.4
4000	40,000	40	5.3
4700	47,000	20–28	6.3
5400	54,000	20–28	7.2
6000	60,000	20–28	8.0
6500	65,000	20–28	8.7
6900	69,000	20–28	9.2
7200	72,000	10–20	9.6
7400	74,000	10–20	9.9
7500	75,000	10–20	10.0

Of concern to the grower is the seeding rate to use to achieve the desired plant population. This information is also given in Table 5.1. The figures are computed on the basis of 15,000 seed per pound and assuming 50% seedling survival of the seeded plants. The latter assumption is about average, but some very large seeded hybrids will contain less than 15,000 seed per pound while others will contain more seed.

PLANTING DEPTH AND DATE

As mentioned in an earlier section, sorghum seed should be planted into moist soil. The depth of planting varies between 1 and 3 inches. Listers are popular in dryland areas because the seed can be placed down into moist soil and still be covered with only 1–3 inches of soil. The lister also helps control surface soil drifting where wind erosion is likely to be a problem.

It is necessary to plant deeper whenever rapid soil-drying conditions are prevalent. An advantage of shallow planting is that soil temperatures are apt to be higher near the surface than at deeper depths.

Dates for planting sorghums vary from February in the southern sections to late June along the colder fringes of the sorghum belt. Sorghum can be grown in combination with winter crops such as wheat, or two sorghum crops may often be grown in one season. Early seeding may be important to avoid grain blasting by hot, dry winds in midsummer and to better fit rainfall patterns and avoid heavy, southwest corn borer and midge populations.

Pest Control

Plant protection programs in sorghum include the control of weeds, diseases, and insects. Annual yield losses from these pests are large. If maximum yields are to be obtained, it is important that the pests be controlled. Those pests likely to result in economic losses by reduced harvestable yields will be discussed.

WEEDS

In the past, weed control in sorghums has primarily been accomplished by mechanical procedures. In recent years, the use of herbicides has played an increasing role. Even today, the use of herbicides only does not solve the weed problem; it is only a supplement to good cultural practices.

Weeds compete with sorghum for moisture, nutrients, and light (Fig. 5.6). Research has shown that only one pigweed per 8 feet of sorghum row may reduce grain yields by 700 pounds per acre. Heavier weed stands reduce yields even more.

Weed control begins with the initial tillage. Tillage operations prior to seeding can help greatly in handling tough perennial weeds. Tillage may also provide the right conditions for germination of annual weeds, which can be killed prior to seeding.

Early weed control is essential for top yields (Fig. 5.7). Because sorghum grows slowly in the seedling stage, weeds find it an easy victim. Cultivation with rotary hoes, knives, disks, and sweeps is used for weed control. Weeds in the row present the greatest chal-

5.6. *Weeds rob sorghum plants of water, nutrients, and light. High-yielding grain sorghum fields are essentially free of weeds.*

5.7. *Early cultivation is necessary in some dryland production systems to control weeds.*

lenge. Timely cultivation, even when sorghums are very young, is often necessary.

Herbicides are grouped according to the time they will be used. Preemergence herbicides are applied as the crop is planted or following planting but before the plants emerge from the soil. The herbicide is often applied simultaneously with the seeding operation with spray nozzles mounted on the planter. Preemergence herbicides may be band applied over the crop row or broadcast over the entire land surface. Preemergence herbicides presently being used include linuron, atrazine, propachlor, metolachlor, and alachlor. Application of metolachlor and alachlor is limited for use on "safened" seed only.

Postemergence herbicides are applied after the crop is up to a stand, that is, until the sorghum is 4–6 inches tall.

The first postemergence herbicide used on sorghum was 2,4-D for the control of broadleaf plants. Its use is continuing, but it must be at the proper stage of growth. It cannot be used without injuring the sorghum until the plants are 4 inches tall, and it must not be used during head development. If the embryonic head can be detected at all by splitting the stalk, it is too late to use 2,4-D; it should not be used until the grain has matured. Other postemergence herbicides being used at present include dicamba, atrazine, bromoxynil, basagran, pendimethalin, trifluralin, and diuron. New materials are being screened each year, and other materials will probably be added annually. Occasionally, pesticides are restricted or suspended. Before any pesticide can be used on sorghum, it must be approved and given final clearance. Rates vary from soil to soil and area to area. Before using herbicides, obtain the latest recommendations and apply according to the manufacturer's label.

Most recommendations for chemicals are made on the basis of pounds of active ingredients per acre and on a broadcast basis. The user should adjust for weight of commercial product needed and for band applications.

Two possible problems need to be considered in the use of herbicides. One is that the amount tolerated by the protected plant can vary according to soil texture, soil moisture, and other climatic conditions. Soils with a low buffering capacity (coarse textured and low organic matter) are especially vulnerable to toxic levels of herbicides for the protected crop. The other problem is that the residual materials may be toxic to subsequent crops (Fig. 5.8).

Seed treatments are now available that will biologically protect the young grain sorghum seedling from injury with the herbicides Lasso and Dual. Monsanto has developed Screen, and Ciba-Geigy has developed Concept II. Several seed companies are currently providing treated (safened) seed to producers who want to use Lasso or Dual for weed control in their grain sorghum (*not* forage sorghums). Producers can also obtain the seed treatment and apply it to their own seed or a hybrid of their choice. Screen was developed specifically for Lasso, while Concept II was developed specifically for Dual. Screen will protect sorghum against both Dual and Lasso injury; likewise, Concept II will protect sorghum from injury by both herbi-

5.8. *This poor stand of grain sorghum was attributed to carryover of a herbicide applied for cotton the previous year.*

cides. The labels for the herbicides and seed treatments should be read closely.

The use of seed treatment safeners has really been a breakthrough for producers. Both Lasso and Dual are very effective for annual grass control. Prior to the development of these safeners, grass control in grain sorghum was very difficult to obtain. In addition to excellent grass control, these two herbicides can provide good control of pigweed and a number of other broadleaf weeds. Yellow nutsedge suppression is also possible with these two herbicides.

It was shown many years ago that cultivation of sorghum was needed only to control weeds. In Table 5.2, the effects of cultivation are clearly demonstrated. If weeds were controlled (where hoeing only was performed), yields were as high as from those plots that were cultivated.

Table 5.2. Effects of cultivation on continuous grain sorghum yields on a Kirkland silt loam soil at Stillwater, Oklahoma

Tillage treatment	Yield
	(*bu/acre*)
No hoeing—no cultivation (weedy)	15.0
Hoeing only	22.9
Shallow cultivation (3 times)	21.6
Shallow cultivation (5 times)	22.1
Shallow cultivation after each rain	22.6
Deep cultivation (3 times)	20.6
Deep cultivation (5 times)	19.7
Deep cultivation after each rain	20.9

Note: Data collected by H. F. Murphy (1931–1940).

For furrow irrigation it may be necessary to clean and reshape water furrows, which at the same time gives some cultivation of the soil.

It is not always possible to completely control weeds with herbicides. For various reasons the applied herbicides may not do a satisfactory job. Weather conditions and soil moisture must be right for adequate control. If herbicides fail, cultivation may be necessary. Give the herbicides a chance, but do not wait too long to use the cultivator. The important thing is to control the weeds.

A combination of practices is required for controlling weeds in sorghums. A combination of tillage, narrow-row spacings, mechanical cultivation, and herbicides gives more dependable weed control than any one method alone. The number and combination of practices needed for any given field for a specific season depend upon many factors including weed species, soil moisture, climatic conditions, and other cultural practices.

WEEDY SORGHUMS

Every producer desires uniform hybrid sorghums and dislikes plants that are taller or obviously nonhybrid. As explained in Chapter 4, changing the plant to a cross-pollinated sorghum naturally allows for outcrossing with other sorghums besides the intended pollinator. These crosses may come from pollen as far away as 5 miles, considerably beyond the normal isolation standards for either certified or commercially produced seed with often more severe requirements. In addition, off-types may be volunteer plants from seed present in the soil. These weedy sorghums result from a dormancy such as with black amber shattercane or because overwintering did not rot or germinate and subsequently destroy the new seedling by frost or cultural practices.

If off-types are from planted seed, they will be uniformly distributed in the field. Also, these weedy sorghums will be in the seed furrow with the intended grain sorghum. To the contrary, volunteer sorghum will be found in clusters or with severe conditions essentially in solid stands.

Clark and Rosenow classify these weedy sorghums into five categories:

1. Tall mutants. These plants resemble the expected hybrid but are 1–2 feet taller. They occur from genetic change of the height genes from either parent. Although of relatively low frequency and having little effect on production, they do give fields a nonuniform appearance. An excessive number suggests failure to use uniform or pure foundation seed or ineffective rogueing during seed production.

2. Off-types or off-color heads. Sorghums that fail to resemble the intended hybrid in color or head shape are likely caused by other grain sorghum pollen and not the male of the hybrid. This pollen may be from nearby commercial sorghum fields or from fertile plants in the male-sterile used in production. The latter off-type will frequently resemble the female parent and could be of somewhat contrasting color, e.g., white or yellow in a red or bronze hybrid, respectively.

3. Silage types. These are tall, vigorous plants with coarse stems. Head types can vary from open to compact and maturity may be similar to the grain sorghum but may be much later. Obviously, these off-types are from silage sorghum pollen. They are more objectionable than the previous categories because of losses during combining. These silage types should be removed before seed are formed to prevent volunteers in succeeding crops.

4. Rhizomatous grassy types. These off-type hybrids are taller than the grain sorghums, plus they tiller prolifically. The heads are open like johnsongrass but produce few or no seed because of genetic (chromosome numbers) differences that make the hybrid highly sterile. Compared to johnsongrass, these plants have short,

weak rhizomes that are more prone to winterkilling following deep-plowing in the fall. These plants should be removed or rogued from any field in which they are found.

5. Nonrhizomatous grassy types. These plants, commonly referred to as shattercane, are heavy tillering with loose, open heads. The most troublesome of this group has easily shattering seed covered with long glumes that have shown dormancy for up to 13 years. Sources of this type rogue are planting seed, acreage previously in sorghum sudangrass, cattle and bird droppings, flood and irrigation water, harvest equipment, and use of Amber or Honey sorgos in the 1920s and 1930s to provide livestock feed, especially during the more droughty years. Plants of this type should be removed within 7 days after flowering, as with category 4.

To control the shattercane-type sorghums that may occur in categories 3 through 5, follow these steps:

1. Identification. Learn to recognize the problem weedy types.

2. Cultural practices. Hand rogue to avoid a much more costly problem in future years. Also, consider rotations with legumes or small grain. Clean machinery before leaving infested fields. Avoid grazing. Severely infested fields should be ensiled.

3. Chemical control. With corn, the use of herbicides plus cultivation normally provides protection. Safened sorghum will allow for the use of "hotter" herbicides to protect against seedling johnsongrass and the nonsafened sorghum off-types.

4. Plant clean seed. Companies make a great and expensive effort to produce seed free of these off-types. Use your experience with companies dedicated to providing the cleanest possible planting seed. Remember, the cheap seed you buy this year, if poorly produced, can be your most expensive long-term investment.

In summary, hybrid sorghums have greatly improved the crop but do carry with them the disadvantage of more off-type or weedy sorghums. Continuous rogueing is highly recommended and justified. Also, consider rotations and appropriate chemical control.

INSECTS

It is estimated that insects reduce the national sorghum crop by 9% annually. In some cases, a complete crop failure is caused by these pests. There are nine important insect enemies of the sorghum crop: greenbug, corn earworm, chinch bug, sorghum midge, cutworm, armyworm, corn leaf aphid, sorghum webworm, and southwestern corn borer (Fig. 5.9).

The *greenbug*, long a menace to wheat and other small grains, has in the last decade become a threat to sorghum production. Ini-

5.9. *Insects that attack grain sorghum include (A) greenbugs, (B) corn earworms, (C) chinch bugs, (D) army-worms, (E) corn leaf aphids, and (F) webworms. (Courtesy Coppock, Tucker, and Maunders)*

A

B

C

D

E

F

tially, greenbugs were noted feeding upon seedlings only, but they now attack even mature sorghums over a wide area of the sorghum belt. Greenbugs not only adapted well to sorghums but also learned to thrive during the hot summers. The sorghum greenbug is now called biotype C or E to distinguish it from the cool-season types. These biotypes are a little lighter in color than other greenbugs. The greenbug, an aphid, feeds upon the cell sap of young, tender leaves and injects a toxin into the plant, making this pest especially injurious. The leaves turn an off-yellow to orange and the plant is badly stunted. The aphids also produce a sticky honeydew substance. The first sign of damage is a wilted plant.

Corn earworms cause extensive damage to sorghum plants. It has been estimated that this pest consumes more than 4% of the U.S. crop. The earworm feeds on newly emerging sorghum heads. It prefers feeding on compact heads and does not damage loose heads nearly as much. The moths lay eggs in the head, which emerge as caterpillars, and do the damage. The distinguishing characteristics of this species are the distinct, short, sharp microspines, resembling whiskers, that are present between the longer hairs on the back. The body color varies from light to dark green, pink, or brownish yellow. When fully grown, the larvae measure up to 1.5 inches in length.

Chinch bugs are sucking insects that feed on the cell sap and starve the sorghum plants. Chinch bugs are black, about 0.1 inch long when full grown, with white wings folded on their backs to form an X. When newly hatched, they are very small, red, and wingless. As they mature, wings develop and the red color disappears. Usually chinch bugs migrate from small grains to sorghum fields after the small grain matures. One control method is to place chemical barriers along field borders. Chinch bugs are most severe during high temperatures and whenever soil moisture supplies are inadequate. Chinch bug damage is first detected by wilting of the lower leaves. In severe cases, the plant turns white and finally dies.

The *sorghum midge* is particularly severe in some areas. These very small, grayish, maggotlike larvae establish themselves close to the developing grain after hatching and from it extract their food. When feeding begins, the larvae turn pinkish in color, which deepens with growth until they are a distinct red. They are found only in the spikelets. An infestation of one larva per spikelet is sufficient to cause complete loss of grain. The adult midge is an orange-colored gnat approximately one-twelfth inch in length. The adult deposits 50–250 eggs; from these, the larvae develop and feed on the developing grain. Egg laying and damage occur only at head flowering time. Evidence of midge damage is blasting of heads (unfilled grain). Cultural practices to help reduce the problem include early, uniform planting within an area and control of johnsongrass, which serves as an alternate host plant.

Cutworms lay eggs in grass, alfalfa, and clovers. Eggs hatch in a few days and the larvae feed on roots. Larvae overwinter about half grown. The young worms may feed on sorghum leaves, but the larger worms feed below the ground. They can completely sever young plants and eat the center of the stalk base on larger plants. Plants attacked by cutworms often suddenly wilt and die. Cutworms work best in cool, moist soil. No treatment is fully effective.

Armyworms feed on leaves, especially in the whorl. The body varies from greenish brown to black, with a narrow light stripe down the middle of the back and four longitudinal stripes on each side. The first three stripes are side by side. The first stripe is mottled brown and darker at the edges. The second is an orange or brown band edged with white. The third stripe is lower on the side and is pale orange edged with white. Armyworms measure up to 1.3 inches in length. They derive their name by traveling in groups.

Corn leaf aphid populations build up on small grains and grasses and then migrate to sorghum. Their direct damage is not as severe as other pests, but they may possibly be carriers of maize dwarf mosaic from johnsongrass. These aphids are tiny, bluish green or gray, and appear in great numbers in the leaf whorl, upper leaves, and head. The usual control is by predators.

Sorghum webworm larvae are small, sluggish caterpillars, with flattened bodies thickly covered with spines and hairs. They are greenish in color and marked with four red to brown longitudinal stripes. They feed upon the ripened grain. The contents of the individual kernels may be partially or completely consumed, leaving only the outside hull intact. In addition to insecticide control, measures include a thorough cleanup of crop residues to destroy the worms that overwinter and planting early to mature the crop before injurious infestations occur. Dry weather coupled with high temperatures often alleviates the infestations.

The *southwestern corn borer* is a white larva with a black head measuring up to 1 inch in length. It feeds in the bud of unfolded leaves. When the leaves open they are then riddled with holes. The larvae have also been known to girdle stalks, causing the plants to lodge. Cultural practices that help reduce the populations are leaving the stubble through the winter and fall disking or shredding to expose the larvae to the weather. In the southern part of the grain sorghum belt, entomologists recommend early planting to evade the borer's heavy population.

Other insects that damage sorghums include wireworms, grasshoppers, banks grass mites (Fig. 5.10), spider mites (Fig. 5.11), the lesser cornstalk borer, and other species of stem borers. These insects normally cause minor damage, but heavy infestations occur periodically and create a problem.

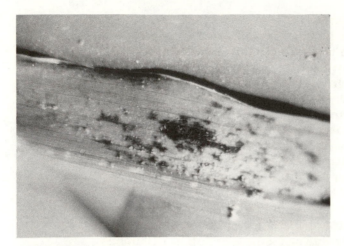

5.10. *Banks grass mites were a problem for grain sorghum 10–15 years ago until resistant varieties were developed.*

5.11. *In the Southwest, occasionally spider mite populations build up to cause damage to a head of grain.*

DISEASES

Disease organisms that attack sorghum are usually classified into four general types by the plant parts affected, those that (1) rot the seed or kill the seedlings, (2) attack the leaves, (3) attack the head and prevent the normal formation of grain, and (4) cause root or stalk rots and prevent normal development of the plant or cause it to break over and fall down before or after maturity.

Root rots and *seedling diseases* of sorghums are primarily caused by fungi. Several species, such as *Fusarium aspergillus, Rhizopus, Penicillium,* and *Helminthosporium,* can cause damage. The fungi invade and destroy the endosperm, causing retarded growth or even seedling death. Problems with fungi attack are most severe when the soil is cold. The fungi thrive at temperatures below 68°F, which is too cold for rapid sorghum growth. The struggling seedlings therefore are easily attacked by the soil fungi.

Seed treatments with fungicides are quite helpful in controlling fungus attack. Planting high-quality seed also helps. The most effective control is planting after the soil is warm enough for rapid germination and seedling establishment.

Diseases attacking sorghum leaves are caused by either fungi or bacteria. Bacterial leaf diseases are quite common but do not generally cause serious problems. The most severe damage occurs whenever a large part of the leaf surface is infected, which may result in improper seed fill. Dead tissue in sorghum becomes pigmented; thereby the sorghum freely shows its leaf infections. The three common leaf diseases of sorghum are bacterial stripe, bacterial streak, and bacterial spot.

Fungi on leaves seldom cause much damage, but occasionally populations build to damaging proportions. Fungus leaf spots can be distinguished from bacterial leaf damage because fungus leaf spots usually are roughened due to the presence of fungal fruiting bodies. Sorghum fungus leaf diseases include leaf blight, zonate leaf spot, gray leaf spot, target spot, sooty stripe, and rust. Most of these diseases occur in humid areas and are not generally a problem in the drier grain sorghum–production areas.

Another leaf disease, anthracnose, is caused by a fungus that also causes stalks to rot. This disease is confined to the more humid areas. The leaves are seldom affected until about the middle of the growing season, at which time small tan to reddish-purple spots appear. The leaf midrib is often strikingly discolored. These spores are spread by wind and rain to other leaves. Resistant sorghums have been developed, but the races are constantly changing.

Downy mildew is a fungus disease infecting sorghums. Symptoms usually appear at the 3- or 4-leaf stage. Small white patches with fluffy spore sacs appear on the underside of leaves as new leaves develop; veins are yellow or white, resembling virus diseases. One symptom of downy mildew that frequently occurs in low wet areas is known as crazy top. The plants become stunted and chlorotic, with thickened, twisted leaves. If heads are produced, they are proliferated.

Disease organisms that attack the heads and prevent seed formation are primarily responsible for smuts. The three smuts of sorghums are covered kernel smut, loose kernel smut, and head smut (Fig. 5.12). Covered kernel smut attacks kernels, and smut spore masses replace the kernel. These spore masses, shaped like the kernel, are enclosed in elongated sacs that are tough and usually do not break as the grain ripens. When smut-contaminated seed are planted, the spores germinate at the same time as the seed. The fungus invades the seedling and grows within the plant unnoticed but becomes apparent when the head emerges. Covered kernel smut is easily controlled by fungicidal treatment of seed.

Loose kernel smut is much less common than covered kernel

5.12. *Several fungal diseases that damage grain sorghum are (A) sorghum head smut, (B) loose kernel smut, and (C) covered kernel smut. (Courtesy Ervin Williams, Oklahoma State University)*

A

B

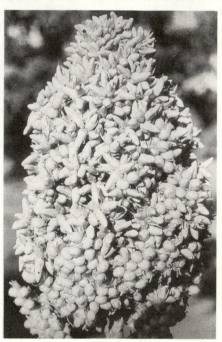

C

smut. It reacts almost identically to the covered kernel smut in growth habits but causes more damage to the plant. Infected plants are often stunted and side branching is induced. This organism is also responsible for secondary infections of other fungi. Loose kernel smut is controlled by seed treatment and use of resistant hybrids.

Head smuts are the most common and destructive of all the smuts. Some of the best hybrids are susceptible. Smut destroys the entire head and is very conspicuous. Damage is apparent as the head emerges. The large gall is covered with a whitish membrane as it emerges from the boot, but it soon breaks and allows the spores to scatter. Unlike other smuts, the head smut spores carry over in the soil for many years. Once head smut organisms are in the soil, seed treatment is not effective, but using a fungicide on the seed prevents infesting clean soil. The only feasible method of control is the use of resistant hybrids.

The most serious *root and stalk diseases* of sorghum are *Periconia* root rot, weak-neck, and stalk rot. The last two do not exhibit visual symptoms until sorghum plants are mature or approaching maturity.

Periconia circinata fungi attack the roots of sorghum. The resultant disease has been referred to as milo disease because it is most serious on milo and its derivatives. This disease has practically been eliminated by resistant hybrids.

Weak-neck used to be a serious problem. This disease causes the heads to break over at the base of the peduncle and then fall to the ground. Hybrids of today seldom have the problem.

Stalk rots are caused by several fungi. The four most common stalk rots are charcoal rot, *Fusarium* stalk rot, *Rhizoctonia* stalk rot, and *Colletotrichum* stalk rot.

Charcoal rot is the most destructive of all sorghum diseases (Fig. 5.13). Symptoms are seldom noted until the plants are near maturity. In severe cases a large majority of the plants lodge. The occur-

5.13. *Charcoal rot causes a disintegration of pith in the stem and leaves the separated fibers (A); hence the weakened stalk breaks and an entire field can be ruined (B). (Courtesy Ervin Williams, Oklahoma State University)*

A   B

rence of the disease is associated with soil and weather conditions that subject the crop to extreme heat or drought during the fruiting period. A positive diagnosis of the disease is made if the diseased stalks become soft at the base, the pith disintegrates, and the separated vascular fibers have a shredded appearance. After a time, the vascular fibers in the collapsed area of the stem become covered with small black sclerotia that look like black pepper or charcoal dust, giving the disease its name. The organism is soil borne. Charcoal rot is most severe under dryland conditions; it is a great threat to sorghum production whenever moisture and climatic conditions are favorable. Most commercial hybrids are susceptible to the disease. There are some tolerant lines used in breeding hybrids, but no hybrids are completely immune.

Fusarium stalk rot is very similar to charcoal rot. The major difference is that the black fruiting bodies are not present. This disease can also cause severe lodging and is aggravated by drought.

Rhizoctonia stalk rot of sorghum is the same organism that attacks cotton and many other crops. Instead of attacking the vascular bundles as do the *Fusarium* and charcoal rot organisms, the *Rhizoctonia* attacks the pith. Large brown fruiting bodies are present on stalk exteriors under leaf sheaths.

The same fungus that causes anthracnose on leaves also is responsible for *Collectotrichum* stalk rot. After infecting the leaves the fungus enters the stalk and spreads throughout the plant. It reduces the movement of plant sap and results in undeveloped heads and seeds. Diseased stalks lodge, breaking at the base.

The only control of stalk rots is by using tolerant or resistant varieties. More breeding work is needed to develop resistant hybrids.

In addition to the diseases discussed, some virus organisms also attack sorghum. Maize dwarf mosaic shows up from 4 to 5 weeks after planting as a faint mottling or interveinal mosaic on upper leaves. The virus is spread by aphids from johnsongrass, which serves as an overwintering host. The best control method is johnsongrass control.

BIRD DAMAGE

In some areas, particularly where sorghum fields are small and interspersed with open grasslands and trees, birds often consume large amounts of grain sorghum. Most bird damage is done at the soft-dough stage when the sugar content of the kernel is high. Birds start eating the grain in the top of the head and work their way downward whenever the food supply is limited.

"Barn" or "city" sparrows cause the most damage because it is difficult to scare them away. Blackbirds are a menace, but they can be scared away with automatic cyanide guns. In some areas, starlings are becoming an increasing problem; they invade sorghum-producing areas in the winter, usually after the sorghum harvest.

Several methods have been tried for repelling birds, but none have really worked satisfactorily. Avoiding bird problems is easier than trying to repel the birds. Cultural practices such as changing planting and harvesting dates are often effective, but selecting bird-resistant hybrids is often a more feasible alternative. Hybrids have been developed with a bitter seed coat that is unpalatable to birds. The bitterness is caused by a tannin that disappears after the kernel ripens. The tannin is present in the bitter form during the dough stage. Be aware, however, that these sorghums are discounted heavily in the marketplace because of their reduced feed value.

6

Fertilization of Grain Sorghum for Profit

General Need

Yields of any crop will usually be limited by some single factor that affects production. In most of the grain sorghum–producing areas, the factor most often limiting production is moisture or soil fertility. Practices such as variety selection, fertilizer use, and seeding rates are usually geared to the anticipated moisture level. For example, if normal available moisture is sufficient to produce 2000 pounds of grain per acre, seeding rate would probably be around 2.5 pounds per acre, variety should be adapted to low-moisture conditions, and fertilizer normally would not be used.

When moisture is adequate or when crops are irrigated, yield potential would be higher and therefore planting rate should be higher (6–10 pounds per acre), a high-yielding hybrid would be used, and the fertilizer rate would be relatively high.

Yield potential is the maximum yield possible when all available knowledge and techniques are ideally applied under a given set of conditions. All controllable factors should be adjusted to yield potential.

All factors affecting production will influence yield potential (Fig. 6.1). To start with, the soil will have an inherent capability to reach a certain yield potential with a given set of climatic conditions. Physical condition of the soil and the resulting soil-moisture-air relationships as well as chemical characteristics are the major soil factors determining yield potential. Water-holding capacity of the soil is of considerable importance in low-rainfall regions. Other factors include varieties and hybrids; rotations and cropping systems; available water; tillage; seeding rate and date of planting; weed, disease, and insect control; and fertilization program.

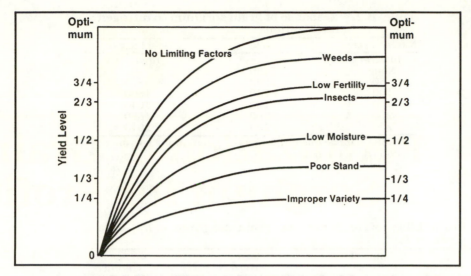

6.1. *Yield potential will be determined by limitation of moisture, fertilizer, weeds, or similar factors influencing growth.*

Fertilizer use must be related to all other practices. It would be unwise to fertilize for an 8000-pound grain yield if expected moisture will limit yield to 4000 pounds; but if a field has the capability for economic production of 8000 pounds of grain, sufficient fertilizer should be applied to produce it.

An example of this is given in Table 6.1. The application of the 80+60+0 fertilizer with two irrigations was not profitable with only 138 pounds per acre extra grain yield. However, the application of 80+60+0 with four irrigations was highly profitable (965 pounds extra yield). This illustrates the interaction between two production factors and the need to tie these together to fit field potential.

Table 6.1. **Relationship between fertilizer use and irrigation on sorghum**

Time and number of irrigations	Yield 0-0-0	Pounds/acre 80+60+0	Yield increase due to fertilizer
Preplant and preboot—2	5040	5178	138
Preplant, preboot, and milk—3	5325	5775	450
Preplant, preboot, flower, and milk—4	6285	7250	965

Note: Data from High Plains Research Foundation, Plainview, Texas.

A quick look at the usual responses to fertilizer by grain sorghum shows that there is a wide range of profitable use of fertilizer. Responses given in Table 6.2 and the economic analysis in Table 6.3 show this wide range of profitability. However, there is one rate of fertilizer that will give the most profit per acre.

Table 6.2. Response of grain sorghum to nitrogen fertilization

Nitrogen rate	Yield	Yield increase
(lb/acre)	(lb/acre)	(lb/acre)
None	5480	. . .
50	6290	810
100	6780	1300
150	7120	1640
200	7360	1880
250	7490	2010

Note: Data adapted from Texas Agricultural Experiment Station trials, Lubbock.

Table 6.3. Profitability of responses on grain sorghum using data given in Table 6.2.

Nitrogen rate	Value of increase*	Cost of fertilizer†	Profit
(lb/acre)			($/acre)
50	40.50	12.00	28.50
100	65.00	22.00	43.00
150	82.00	32.00	50.00
200	94.00	42.00	52.00
250	100.50	52.00	48.50

*Value of grain sorghum computed at $5.00/cwt.
†Cost of fertilizer computed at $0.20/lb nitrogen plus $2.00/acre application costs.

The optimum fertilization rate is the one that will give the maximum net profit per acre. In this case the rate is around 200 pounds nitrogen per acre according to Table 6.3. A higher yield was obtained with 250 pounds nitrogen, but it was not as profitable. The highest yield possible will usually produce less profit per acre than a yield that is slightly less than maximum.

An illustration of how profits vary with a fertilizer rate and soil test nitrogen level is shown in Figure 6.2. Rates of nitrogen may be profitable from very low to very high if the soil test is low. However, the maximum profit from fertilizer, as shown in the example, would be at about the 90 pounds per acre rate for a low soil test value. For a medium nitrogen soil test, the maximum profit rate would be about 60 pounds of nitrogen per acre.

What happens to the profitability of a fertilizer rate if the price of grain sorghum goes up or the cost of fertilizer goes down? This is illustrated in Table 6.4, which shows that more should be used. Note that in the table the rate of nitrogen would vary plus or minus 20 pounds per acre for each 1 cent increase or decrease in the price of a pound of nitrogen. It is also noteworthy that for each additional 10 cents per hundredweight received for grain sorghum, the profitable nitrogen rate would increase by about 4 pounds per acre at a given price for nitrogen. If the reverse should occur, that is, a lower price

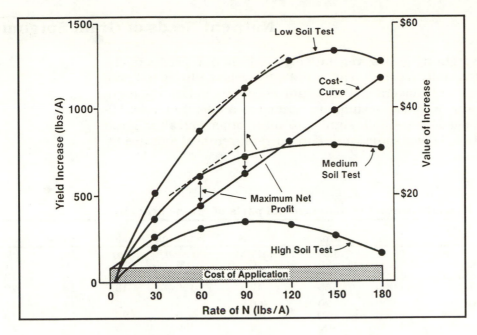

6.2. *As fertilizer rates increase, yields will also increase if the fertilizer is needed, but a maximum profit fertilizer rate is reached before maximum yield is obtained.*

for grain sorghum or a higher cost of fertilizer, lower rates should be used to maximize profits.

Fertilization and any other practices that increase yield will decrease the unit costs of producing grain sorghum; certain costs will be fixed costs. These include items such as taxes, land costs, plowing, planting, herbicide application, labor, and management. Fixed costs per unit of grain sorghum will vary with yield. If the fixed costs are $60 per acre, a yield of 3000 pounds would give a $2 cost per hundredweight, whereas, if the yield were 6000 pounds, cost per hundredweight would be only $1. It is generally desirable to produce the optimum yields in order to reduce the fixed costs per unit of production.

The variable costs such as fertilizer, seed, water, and herbicide have to be adjusted to yield potential and used at optimum rates to maximize profits and minimize the fixed costs per unit of production.

Table 6.4. **Relationship between price of grain sorghum, cost of nitrogen, and nitrogen rate in pounds per acre**

	Price of grain sorghum ($/cwt)		
Cost of nitrogen (*cents/lb*)	$5	$6	$7
15	200	240	280
18	140	180	220
21	80	120	160
24	20	60	100

Nutrient Needs of Grain Sorghum

Grain sorghum grows vegetatively and then reproductively. Such plants usually respond readily to high levels of nitrogen. Phosphorus needs are moderate. Potassium needs are relatively high. Sulfur, calcium, and magnesium are needed at moderate levels. Micronutrient needs are small, similar to most other cereal or grain plants. Table 6.5 gives some values on nutrient removal and use by grain sorghum.

Table 6.5. Nutrient content of aboveground parts of grain sorghum

Yield	Plant part	Dry* matter	Nitrogen	Phos-phorus	Potassium	Calcium	Sulfur	Mag-nesium
(lb/acre)		(lb)	(lb)	(lbP₂O₅)	(lb K₂O)	(lb)	(lb)	(lb)
				$(lb\,P_2O_5)$	$(lb\,K_2O)$			
6,000	Grain	5,100	95	30	25	3	10	7
	Stover	6,800	70	12	100	15	8	13
	Total	11,900	165	42	125	18	18	20
8,000	Grain	6,800	120	60	30	5	14	10
	Stover	8,000	80	16	120	20	12	15
	Total	14,800	200	76	150	25	26	25
10,000	Grain	8,500	145	70	35	7	18	13
	Stover	9,500	95	20	135	25	20	18
	Total	18,000	240	90	170	32	38	31

Note: Total removal for micronutrients at an 8000-pound level is approximately: iron, 1.5 lb; zinc, 0.8 lb; manganese, 1.2 lb; copper, 0.1 lb; boron, 1.5 lb.

*Dry matter based on assumption of 15% moisture in grain. Stover tonnage will usually exceed yield by a greater quantity at lower yields than at higher yields.

Grain sorghum will usually respond to nitrogen, provided yield potential is at least 2500–3000 pounds per acre. A summary reported by Valentine and Onken shows that nitrogen responses were obtained in 85% of the experiments (84 tests over a 9-year period) conducted in the Texas high plains, an 18-inch annual rainfall area often supplemented by irrigation water. Soil in most grain sorghum areas are often deficient in phosphorus and will usually respond to phosphorus even through responses may not be large. Continuous phosphorus use over a long period can result in a soil phosphorus buildup and will often minimize phosphorus fertilizer needs. Figure 6.3 gives some typical responses to nitrogen and phosphorus on sandy loam soils.

Soils in most of the grain sorghum area are usually well supplied with potassium, calcium, and magnesium. Soils in the higher rainfall areas may be deficient in calcium and magnesium, and these elements may need to be applied. Sulfur may be beneficial in certain parts of the growing area.

Micronutrients are needed whenever the soil is unable to supply them to the plants in sufficient quantities. The quantity of micronu-

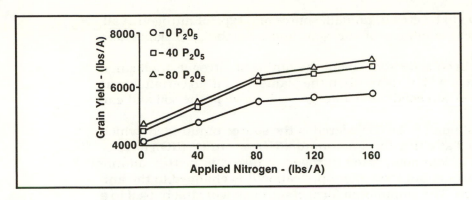

6.3 *Typical grain sorghum yield response to nitrogen and phosphorus on fine sandy loam soils in the United States Great Plains area. (Courtesy Texas Agricultural Experiment Station)*

trients present in the soil is usually adequate for high yields, but the availability may be low due to high soil pH. Micronutrients should be applied whenever a known deficiency exists. Thereafter, maintenance levels can be included in the fertilizer to insure that deficiencies do not develop again in the future.

Fertilizing with Macronutrients

NITROGEN

The rate of nitrogen to use can often be estimated with a fair degree of accuracy. A general guideline is given in Table 6.6, which takes into account the yield potential and the fact that the first increments of nitrogen applied will be more efficiently used than additional amounts. These rates will need to be changed if losses due to leaching are anticipated. The suggested rates are based on average conditions, with other production factors adjusted accordingly.

Table 6.6. Nitrogen rates for various grain sorghum yield levels

Yield potential	Pounds nitrogen needed/100 lb increase in grain	Pounds nitrogen needed/1000 lb increase in grain	Total pounds nitrogen needed
2000–3000	. . .	. . .	None
3000–4000	2	20	20
4000–5000	3	30	50
5000–6000	4	40	90
6000–7000	5	50	140
7000–8000	6	60	200
8000+	6+	60+	200+

A number of factors to consider concerning when and how to apply nitrogen and what kind to use include soil temperature, soil texture, crop to be grown, rainfall, source and type of nitrogen available, physical condition of the soil, and availability of labor and equipment.

The main consideration about the timing of nitrogen application is being sure it will be there when the plants need it. In certain forms and under certain conditions, nitrogen will move in the soil and can be lost.

The first point to be considered is the source of nitrogen, which is categorized as either the ammonium (NH_4^+) or nitrate (NO_3^-) form or both. Urea, a commonly used nitrogen fertilizer, is neither ammonium nor nitrate, but soon after application it is changed to the ammonium form. The ammonium form of nitrogen will attach itself to a clay particle and resist leaching. The nitrate form will not become attached and will move with the soil moisture.

Nitrate nitrogen moves downward as the soil moisture moves downward after an irrigation or rainfall. As the soil dries and moisture moves upward by capillarity, the nitrates move upward with it. In irrigated soils, white salts can be seen on the surface of beds after preirrigation. These are often nitrate salts.

Soil temperature is important because most of the nitrogen in the ammonium form will stay in that form if soil temperatures are below 45–55°F. Above that temperature, provided there is adequate but not excessive moisture and other conditions are right, the ammonium form will convert to the nitrate form, which will then move with soil moisture.

Texture has a twofold effect. Finer textured soils will hold the ammonium ion more effectively. However, most soils, except very sandy soils, have enough clay in them to hold a relatively high rate of an ammonium form of nitrogen. The texture of the soil will also influence the speed and distance that moisture moves through the soil. It will move less rapidly and deeply through a clay soil and, consequently, nitrate nitrogen will be less apt to leach on fine-textured soils. Movement can be quite rapid and deep through a sandy soil as water moves downward.

Rainfall will obviously influence nitrate movement. If it is limited, loss of nitrates are rare simply because there is not enough moisture to move it out of the root zone. This is true of a large portion of the sorghum-producing areas. Occasional rainy periods, however, will occur with sufficient rainfall to move nitrates out of the root zone.

With the exception of sandy soils and areas of high rainfall, nitrogen can generally be applied when it is convenient and when labor and equipment are available. In sandier soil areas and with high rainfall, it is generally desirable to delay applications of nitrogen fertilizer until close to the times of plant need or use split applications.

Considerable research has been done to compare various times of application of anhydrous ammonia. Data typical of most of that research, illustrated in Figure 6.4, show that nitrogen can be applied almost anytime on the finer textured clay and clay loam soil types. Applications in coarse-textured soils should be delayed until just prior to planting or an early sidedress. Data shown in Figure 6.5 suggest that anytime after preplant irrigation should be satisfactory on sandy soils. As mentioned previously, it depends on the amount of water that will be moving through the soil and the soil temperature.

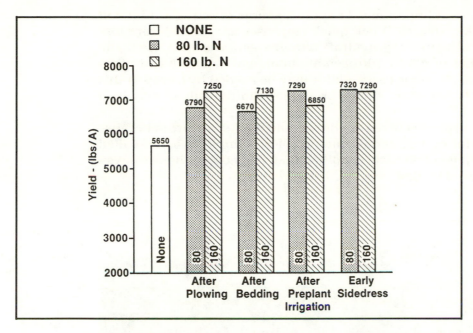

6.4. *On fine-textured clay loam soils, nitrogen can be applied at almost any time with equal effectiveness.*

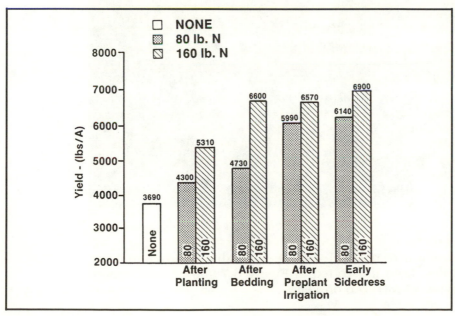

6.5. *On coarse-textured, fine sandy loam soils, consideration should be given to later applications.*

Nitrogen fertilizers are applied as a gas, liquid, or solid. They can be applied as a broadcast or topdressed application, chiseled or banded into the soil, applied in the irrigation water, or applied by any number of different methods. Type of nitrogen used as well as soil and crop characteristics will determine the best method.

The best source of nitrogen to use depends on agronomic considerations, physical form of the nitrogen available, soil conditions, and cost.

It may be of importance to apply a source of nitrogen that will give a quick response. When quick responses are needed from top-dressing or side-dressing, a nitrate nitrogen source is often used. The major sources of nitrate nitrogen are ammonium nitrate and nitrogen solutions. In some cases, sulfur may be needed by a crop. If this is the case, ammonium sulfate or ammonium thiosulfate may be a good choice.

Some nitrogen sources are better suited for some areas because of physical form. Anhydrous ammonia, for example, is better suited for soils with good physical condition than for those with poor physical condition (Fig. 6.6).

6.6. *Anhydrous ammonia is one of the principal sources of nitrogen used for grain sorghum. (Courtesy Tyler Industries)*

A large amount of research has been conducted in comparing nitrogen sources. Data illustrated in Figure 6.7 show the effects of the various types of nitrogen on yield.

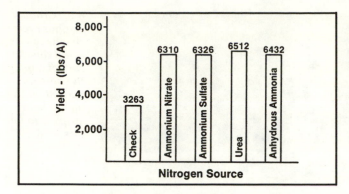

6.7. *Differences in response to the various sources of nitrogen are minimal. (Courtesy Texas Agricultural Experiment Station)*

Fertilizer will carry over from one year to the next. In the year of application, grain sorghum, for example, will normally utilize 40–70% of the nitrogen applied. The utilization rate for phosphorus is around 15–20%. For potassium, it is 30–40%. This means that continual applications will result in an accumulation of these nutrients. The carryover effect can be significant. Suggested nitrogen rates are listed in Table 6.6.

How long will nitrogen have a residual effect? Usually for no more than 1 year, and then noticeable decreases in yield may occur. Typical data to illustrate this comes from a 7-year study at the Southwestern Great Plains Field Station in the Texas Panhandle (Fig. 6.8). The yield without nitrogen generally decreased each year. With applied nitrogen, yields ranged around 6000 pounds per acre, depending on the year. When no nitrogen was applied, yields dropped rapidly. They continued to drop significantly when no nitro-

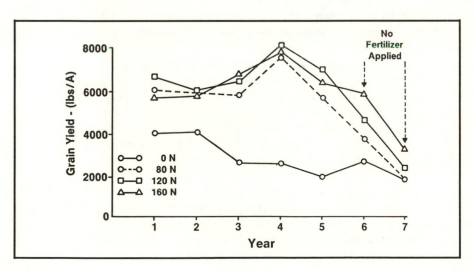

6.8. *Data illustrating response to nitrogen and the carryover effect of nitrogen. (Courtesy Texas Agricultural Experiment Station)*

6.9. *Grain sorghum responds to nitrogen if the soil cannot supply the amount needed as shown by the light strips through this field.*

gen was applied in the following year. Nitrogen must be applied each year to maintain yields unless there is above-normal carryover (Fig. 6.9).

PHOSPHORUS

Phosphorus rates are more difficult to predict, because of differences in soils, the manner in which phosphorus reacts in the soil, and the many different types of phosphorus fertilizers. Determine need by soil tests.

Time of application of phosphorus is less critical than for nitrogen. Phosphorus will undergo chemical reactions and change to a less available form with time. The rapidity of these reactions depends on temperature, moisture, soil texture, soil pH, and type of phosphorus fertilizer used. Phosphorus reactions may be complete on some soils within 3–4 weeks after it is applied and will reach about the same state in the reversion process whether it is applied in the fall or late spring. It will stay in the reacted form until conditions change in the soil or until a plant absorbs some of the available phosphorus. Hence it is not too critical as to when phosphorus is applied for grain sorghum.

The reversion of phosphorus to less soluble forms is important but usually not critical. Many soils in the sorghum-producing area are alkaline and calcareous, which means that a soil contains a high level of calcium or even contains excess calcium as calcium carbonate. This means that phosphorus reversion, often called fixation, does take place, and even though it is usually thought of as being undesirable, the reversion does not negate crop responses.

Phosphorus exists in several forms in the soil. The available forms in the soil solution are the $H_2PO_4^-$ and $HPO_4^=$ ions. The concentration of these ions in the soil solution at a given time is small and must be continuously replenished if plants are to grow at an optimum rate. The relative concentration of the two available phosphate ions is related to soil acidity. The $H_2PO_4^-$ ion predominates in an acid soil, whereas the $HPO_4^=$ ion is more plentiful in alkaline soils. Because the plant requires less energy to absorb the $H_2PO_4^-$ ion than the $HPO_4^=$ ion, higher phosphate levels are usually required in alkaline and calcareous soils.

Phosphorus also exists in different calcium phosphate forms and many different types of phosphorus complexes in the soil. It also is adsorbed on the surfaces of clay particles and may be precipitated on the surface of free calcium carbonate.

The main point of concern is whether the plant absorbs the phosphorus. The plant will absorb the available $H_2PO_4^-$ and $HPO_4^=$ ions from the soil solution. Then some of the less available forms of phosphorus will shift back to an available phosphorus ion that the plant can then utilize. This adjusting to the equilibrium means that phosphorus will become available provided there is an adequate reserve.

Placement of phosphorus fertilizer will be important for grain sorghum in some areas. Early access to phosphorus by the plant is important. Therefore, placement so that the young plant can get to it early is usually desirable. It will be more important in soils that stay cool and wet in the spring or in soils that are extremely low in phosphorus. Banding close to the seed, usually within 2–4 inches, would be desirable for these conditions. Broadcasting normally is suitable for grain sorghum where soil phosphorus is not critically low and where soils warm up quickly in the spring. Grain sorghum has a fibrous root system; hence it can "forage" for phosphorus scattered throughout the area of root penetration. Method of application (Fig. 6.10) of phosphorus is often determined by the source, time of application, and factors other than agronomic considerations.

6.10. *Fertilizer in the form of manure used to be spread by hand from wagons, but today, fertilizer is spread by large equipment with flotation tires.*

The main phosphorus sources used in most grain sorghum–producing areas can be grouped into two categories: calcium phosphates and ammonium phosphates. Information is given on these sources in Table 6.7.

Table 6.7. Common phosphorus fertilizers and their characteristics

Group	Physical state	Examples	Water solubility	Form of phosphate	Chemical formula
			(%)		
Calcium phosphates	Solid	0–20–0	85–90	Ortho.	$Ca(H_2PO_4)_2$
		0–46–0	95–98	Ortho.	"
Ammonium phosphates	Solid	11–48–0	100	Ortho.	$NH_4H_2PO_4$
		18–46–0	100	Ortho.	"
		16–20–0	100	Ortho.	"
		15–62–0	100	Poly.	$(NH_4)_3HP_2O_7$
	Liquid	7–21–0	100	Ortho.	$NH_4H_2PO_4$
		10–34–0	100	Poly.	$(NH_4)_3HP_2O_7$
		11–37–0	100	Poly.	"
Phosphoric acids	Liquid	0–52–0	100	Ortho.	H_3PO_4
		0–54–0	100	Ortho.	H_3PO_4
		0–75–0	100	Poly.	$H_4P_2O_7$
Ammoniated superphosphate	Solid	10–20–0	20–60	Ortho.	Variable
Rock phosphates	Solid	0–1–0	Nil	Ortho.	$Ca_{10}(PO_4)_6X_2$
Nitric phosphates	Solid	20–20–0	10–80	Ortho.	Variable

Criteria used to determine the effectiveness of the phosphorus source are water solubility, particle size, and associated salts. Generally, a phosphorus source with high water solubility is desirable compared to one of lower solubility. Particle size and water solubility are related. If water solubility is low, smaller particle size is usually desirable. If solubility is high, it would generally be desirable to have a larger particle size.

The ion associated with the phosphorus can also be of importance. The one usually associated is either the ammonium or calcium ion. It is generally desirable to have nitrogen and phosphorus associated together in the soil. This will usually result in an increased uptake of phosphorus by the plant. Calcium as an associated ion might be of some benefit in acid soils but would be of little value in neutral to alkaline soils.

The sulfate ion is also present in some phosphorus fertilizers such as ordinary superphosphate (0-20-0); it is present as calcium sulfate. The sulfate ion is available as a plant nutrient; however, it would probably have little effect on phosphorus absorption and uptake in alkaline soils, nor does it have any effect on reducing pH in alkaline soil.

POTASSIUM

Potassium can be applied either broadcast or banded for good results. It can be applied at the same time as the nitrogen and phosphorus.

The principal potassium source used on grain sorghum is potassium chloride (KCl), which usually has an analysis of 60% K_2O (0-0-60 or 0-0-62). In areas where sulfur and/or magnesium are needed in addition to potassium, the use of potassium-magnesium sulfate ($K_2SO_4 \cdot 2MgSO_4$) has proven to be popular. Potassium sulfate (K_2SO_4) and potassium nitrate (KNO_3) are used in some areas as potassium sources (Table 6.8).

Table 6.8. Common sources of potassium and sulfur

Material	Formula	Grade	Percentage K_2O	S
Muriate of potash	KCl	0-0-60	60	. . .
Potassium sulfate	K_2SO_4	0-0-50	50	17
Potassium- magnesium sulfate	$K_2SO_4 \cdot 2MgSO_4$	0-0-22	22	22
Potassium nitrate	KNO_3	13-0-44	44	. . .
Ammonium sulfate	$(NH_4)_2SO_4$	21-0-0	..	23
Gypsum	$CaSO_4 \cdot 2H_2O$	. . .	..	19
Elemental sulfur	S	. . .	..	100
Ammonium thiosulfate	$(NH_4)_2S_2O_3$	12-0-0	..	26
Ammonium polysulfide	NH_4S_x	20-0-0	..	40

SULFUR

Sulfur has given responses on grain sorghum in some areas. Where needed, an application of 1 pound sulfur for each 20 pounds nitrogen is often suggested. Sulfur deficiencies are most apt to occur on soils that are coarse textured and low in organic matter.

The increased interest in sulfur use over the grain sorghum belt has resulted in increased use of several sulfur products. Elemental sulfur is one source. It is now generally available in a prilled or flaked form. Some of these are water degradable, which results in a more rapid breakdown and conversion of the elemental sulfur.

Sulfur is also present in some fertilizers used principally for another nutrient. Some common ones are ammonium sulfate, potassium sulfate, ammonium phosphate sulfate, and ordinary superphosphate (which contains calcium sulfate). A large number of other fertilizers contain small quantities of sulfur.

Sulfur can be applied in elemental form, as a thiosulfate, as a polysulfide, or as sulfate. The elemental, thiosulfate, and polysulfide forms have to undergo transformations by bacteria to the sulfate form. These sulfur sources have an acidifying effect, or an acid residual, after it undergoes this transformation. Conditions for this change to occur are similar to those discussed previously for nitrogen transformations. These include the right range of temperature, proper moisture, adequate nutrients, and soil pH.

Two liquid sulfur sources are thiosulfates and polysulfides. Ammonium thiosulfate (12-0-0-26S) is applied alone or mixed with other liquid fertilizers. Ammonium polysulfide (20-0-0-40S) is usually applied in irrigation water as a direct soil application.

Sulfur, if applied as a plant nutrient, can be applied with other nutrients at the same time using the same method of application. If sulfur is applied as a soil amendment to reduce soil pH, it can be applied at any time it is convenient. The dry forms are usually applied with other fertilizers as a broadcast or banded application.

CALCIUM AND MAGNESIUM

Responses to calcium and magnesium have been reported mostly in the more humid areas where sorghum is grown, principally on acid soils.

These two elements, when needed in acid soil areas, are usually applied as limestone. It will be calcitic limestone (calcium carbonate) where only calcium is needed. If both are needed, dolomitic limestone, which contains both calcium and magnesium carbonate, will be used. These materials are usually broadcast at high rates (½ ton up to 5 tons per acre) and incorporated into the soil.

If magnesium is needed in areas where soils are alkaline, it is usually applied as a potassium-magnesium sulfate or a similar type product. This material is usually mixed with nitrogen-phosphorus fertilizers and is broadcast.

Fertilizing with Micronutrients

Soil conditions that may result in micronutrient deficiencies or decreased availability are usually high pH, low organic matter, or coarse texture. These conditions will not necessarily mean that micronutrients will be deficient, but it may be good to watch for the possibility of a deficiency.

A large percentage of the grain sorghum production is in areas where soils are principally in the neutral to alkaline range, relatively low in organic matter, and often sandy texture. Therefore, micronutrient levels or availability may be low. The micronutrient most often found to be deficient for grain sorghum is iron. Zinc and manganese have increased yields in some areas. Responses to copper, boron, and molybdenum have seldom occurred. Table 6.9 gives the conditions where deficiencies might be apt to occur and the conditions that encourage their reversion to a less available state.

The metallic micronutrients—iron, zinc, manganese, and copper—are usually applied as salts, chelates, or frits.

Table 6.9. Conditions influencing deficiencies and availability of micronutrients

Micronutrient	Deficiencies or low availability	Promoting available state	Promoting fixation to unavailable state
Iron	Coarse textured; high soil pH and CaCO₃; land leveling; high oxidation state of soil; soil compaction	Release by microbial action from organic residues; adsorption on the clay particles of moderately acid soils; formation of soluble chelates	Strongly acid or calcareous soils; high soluble phosphorus iron precipitates as oxide, hydroxide or phosphate insolubles; alkaline soils form insoluble Fe(OH)₃
Zinc	Low total zinc; coarse textured; low organic matter; high pH, free CaCO₃; land leveling	Release by microbial action from organic residues; adsorbed on clay particles in soil; formaton of chelates	High pH: phosphate, carbonate, and clay soils precipitate insoluble zinc as phosphate, calcium, or magnesium salts
Manganese	High pH, calcareous soils; thin, peaty, poorly drained calcareous soils; sandy acid soils	Acid soils; high OM soils; soils high in phosphorus retain divalent manganese as organic or phosphate salts; formation of chelates	Neutral or alkaline soils; highly oxidized soil precipitates manganese as insoluble salts of higher oxide forms
Copper	High pH, calcareous soils; high OM soils such as peat, muck	Release by microbial action from organic resides; formation of soluble chelates	Low OM soils with high pH precipitate insoluble copper hydroxide
Boron	Coarse textured, well-drained soils; high pH and calcium	Release by microbial action from organic residues in acid to neutral soils with high OM content	High OM in alkaline soils (pH 7.5+) causes precipitation of an insoluble calcium-boron-organic complex
Molybdenum	Acid soils	Release by microbial action from organic residues in neutral to alkaline soil; liming acid soils	Acid soils or soils high in iron or aluminum oxides fix molybdenum on metallic oxides

Note: OM = organic matter.

The salt forms are usually sulfates such as iron sulfate (copperas) and zinc sulfate. The carbonate and oxide forms can also be used. The main advantage of these forms is that they are inexpensive and the cost per acre is quite low. The primary disadvantage is that these forms may readily revert to a low-availability state, especially in alkaline calcareous soils.

Chelates are organic molecules that react with metal ions and keep them in a more readily available form. They can be categorized into three groups: synthetic organic molecules; long-chain, natural organic molecules; and short-chain, small organic molecules. The synthetics are the strongest in complexing metal ions and include EDDHA, DTPA, EDTA, and many others. The long-chain group includes lignin sulfonates, humic and fulvic acids, and polyflavinoids. The short-chain group includes, for example, tartaric and citric acids. Per pound of the metal ion, the synthetics are most expensive, the long-chain groups moderate in cost, and the short chains usually the least expensive.

The main advantages of the chelate forms are that the reactivity of the micronutrient is decreased and hence will persist in a relatively high available state, and they are usually soluble in water and in most fertilizer solutions (this varies and should be checked before they are mixed). Chelate forms generally cost more than the sulfate forms. There is a considerable variability in effectiveness from region to region among the chelated forms. Consult your local fertilizer dealer on the ones that work best in your area.

Fritted trace elements are micronutrients that have been fused into a glass form. They are then finely ground and applied in this manner. The principal advantage of "frits" is that they dissolve slowly and are slowly available. Hence their reactivity is decreased and they are slow to revert to an unavailable form.

The source of micronutrients to be chosen depends upon price and effectiveness. Some sources of micronutrients are listed in Table 6.10.

The best form to use may be related to method of application. Micronutrients are well suited for a foliar application. Because of the need for only small quantities, this method is often recommended as a way to remedy micronutrient deficiencies. Both the chelate and salt forms work well as a spray application on the foliage. For foliar applications to be effective, they should be applied so as to obtain thorough coverage. A spreader-sticker added to the solution is usually beneficial. Apply late in the evening or early in the morning for best results. Be careful in spraying micronutrient salts and chelates on foliage. They can burn if too concentrated. Use only the rate suggested and be sure of formulations.

Applications of micronutrients to the soil will usually be made with the application of other fertilizers. They are usually either broadcast and incorporated or banded. Reducing the surface-area contact of the salt forms with the soil is desirable. Of the four metallic micronutrients, iron salts are the least suited for broadcast, whereas zinc salts are the best suited. Copper and manganese salts are intermediate.

Table 6.10. Sources of micronutrients

Element and material	Type	Concentration
		(%)
Iron		
Ferrous sulfate	Salt	20
Synthetic chelates	Chelate	6–12
Organic complexes	Chelate	5–10
Frits	Frits	14
Zinc		
Zinc sulfate	Salt	18–36
Zinc oxide	Salt	70–80
Synthetic chelates	Chelate	10–14
Organic complexes	Chelate	5–12
Frits	Frits	4–7
Manganese		
Manganese sulfate	Salt	27–28
Manganese oxide	Salt	48–60
Synthetic chelates	Chelate	10–12
Organic complexes	Chelate	8–12
Frits	Frits	10–35
Copper		
Copper sulfate	Salt	25
Synthetic chelates	Chelate	13
Organic complexes	Chelate	5–12
Frits	Frits	1–20
Boron		
Borax	Salt	11
Sodium borate	Salt	14–20
Frits	Frits	3–6
Molybdenum		
Sodium molybdate	Salt	40
Ammonium molybdate	Salt	49
Frits	Frits	1–30

When applied with other fertilizers, micronutrients will usually be blended, sprayed on the fertilizer granules, or incorporated into the granule. For soils in which micronutrients stay in an available form, such as acid to neutral soils, any of the above three are satisfactory. However, in alkaline soils, incorporating micronutrients into the granule or on the surface of fertilizer granules should be more effective in general than if they are applied as separate granules.

Be careful in the use of micronutrients. If used improperly, they can have a detrimental effect on yield. Three of the micronutrients most likely to be toxic to plant growth if applied in excess are boron, copper, and manganese. Use these only at recommended rates. Levels of micronutrients for grain sorghum will vary depending on soil conditions.

IRON

Grain sorghum is more susceptible to iron deficiencies than to deficiencies of any of the other micronutrients. A deficiency of iron results in reduced chlorophyll formation, producing a symptom called iron-deficiency chlorosis. It is not too uncommon to see iron-deficiency chlorosis in fields of grain sorghum. The deficiency can be corrected with a foliar spray of ferrous sulfate (often called copperas) (Fig. 6.11). A 2.5% solution of ferrous sulfate consisting of 2 pounds of the material in 10 gallons of water per acre is usually an effective rate if chlorosis is present. If the chlorosis persists, it may be necessary to repeat the application one or two times. It should usually be applied as soon as the chlorosis appears. Response to this type of foliar spray is usually great and requires little expense. One problem is that the chlorosis may affect only a small portion of the field and it may not be economical to treat such a small area. Soil applications of iron sulfate are seldom effective for grain sorghum. The inorganic iron salts revert rapidly to an unavailable form in calcareous soils.

Iron-deficiency chlorosis can also be controlled with an iron chelate. A rate of about 0.1 pound iron per acre should be applied for foliar sprays (this would be 2 pounds of a 5% iron chelate material). Chelated iron at the rate of 1 pound iron per acre is often suggested for acid to neutral soils.

ZINC

Zinc deficiencies in the soil can usually be corrected by a broadcast application of 1–5 pounds of actual zinc per acre in the form of an inorganic salt (3–14 pounds of a 36% zinc sulfate). For chelate sources, apply 0.5–1 pound actual zinc broadcast (8–16 pounds of a 6% zinc chelate). If deficiencies are extremely severe, double these

6.11. *Foliar sprays of iron on grain sorghum will correct iron deficiencies on soils containing excess quantities of calcium carbonate. (Courtesy Texas Agricultural Experiment Station)*

rates. Zinc can be applied with a row fertilizer. It may need to be applied each year until available zinc levels build up; however, 3–5 pounds of zinc in one application should be effective for several years.

MANGANESE

Responses to and need for manganese have not been as general as for iron and zinc. A broadcast soil application of 5–10 pounds of the actual metal per acre is usually effective. The lower rates should be used on the neutral to acid soils and higher rates on the alkaline, calcareous soils. Chelated manganese can be applied at a rate of 1–4 pounds per acre. Banded manganese may be considerably more effective than a broadcast application.

COPPER

Copper deficiencies on grain sorghum are extremely rare. Foliar-applied copper or high levels of available copper in the soil can cause toxicity. Be careful in its use.

OTHER MICRONUTRIENTS

Boron and molybdenum are not expected to be needed in grain sorghum-producing areas. If a need were definitely established, a suggested rate would be 1 pound boron per acre from borax or sodium borate. For molybdenum, the recommended rate is from 1 to 2 grams per acre from sodium molybdate or a similar molybdenum compound. This small amount of molybdenum can best be applied by treating the seed. Be careful in the use of these micronutrients.

Nutrient Balance

The proper balance of nutrients is important to plants. If phosphorus is deficient and nitrogen is applied, the nitrogen may give only a small or even no response. The reverse could be true. If a micronutrient such as iron is deficient, the application of any other essential nutrient would be of little value. An example can be cited on a clay loam soil when the yield without nitrogen and phosphorus was 5300 pounds per acre. An application of 80 pounds of applied nitrogen alone was sufficient to produce about 6500 pounds of grain per acre, or an increase of approximately 1200 pounds per acre. Additional nitrogen rates of 120 or 160 pounds nitrogen per acre gave very little additional increase. However, when applied in combination with 40 pounds P_2O_5 per acre, the additional nitrogen produced over 7400 pounds per acre.

7

Grain Quality and Plant Composition

GRAIN QUALITY is a nebulous term. It means different things to different people. Generally, quality criteria are determined by the use of the product. To the feedlot operator, quality means energy, protein, and minerals. To the broiler producer, it also means vitamins and carotene content. To those that use it for food, quality means carbohydrates and types of protein. To all, therefore, the composition of the grain is of prime importance.

The chemical composition of the vegetative growth controls to a large measure the yield and composition of the grain, and the producer can control this to some extent.

Grain Composition

The grain sorghum kernel is composed primarily of starches, sugars, proteins, fats, and oils. Protein will average about 10%, fat 3%, and starches and sugars about 70%. The seed coat also contains waxes. The rest of the kernel is made up of other miscellaneous constituents and the mineral elements. The average mineral composition of grain sorghum is given in Table 7.1. Some quality compo-

Table 7.1. Average mineral composition of grain sorghum

Element	Content
	(%)
Nitrogen	1.80
Phosphorus	0.30
Potassium	0.40
Sulfur	0.15
Calcium	0.04
Magnesium	0.15
Iron	0.005
Manganese	0.001
Copper	0.001

nents of grain are given in Table 7.2. Grain sorghum also contains many of the vitamins: riboflavin, niacin, and pantothenic acid (B-complex vitamins), which are comparable to other cereal grains.

Table 7.2.   Some quality components of various grains

Component	Sorghum	Corn	Barley	Wheat
Digestible energy (kcal/kg)	3453	3610	3080	3520
Protein (%)	11.0	10.0	11.6	12.7
Lysine (%)	0.27	0.13	0.53	0.45
Methionine + cystine (%)	0.27	0.18	0.36	0.36
Tryptophan (%)	0.09	0.09	0.18	0.18
Calcium (%)	0.04	0.01	0.08	0.05
Phosphorus (%)	0.30	0.31	0.42	0.36
Fiber (%)	2.0	2.0	5.0	3.0
Ether extract (%)	2.8	3.9	1.9	1.7

PROTEIN

Of all the constituents of sorghums, protein content varies most and is one of the most important constituents of feeds and foods. Proteins are made up of the amino acids. The kinds of amino acids present in grain sorghum fulfill the requirements for cattle but are unbalanced for poultry and swine. Grain sorghums are low in the amino acid lycine. Plant breeders are teaming up with biochemists in efforts to improve the content of this amino acid in sorghums.

As yields of grain sorghum increase at a given nitrogen level, the protein content tends to decrease. For a given yield level, the protein content of the grain can be increased only by the uptake of additional nitrogen by the plant. The relationship between fertilizer nitrogen and protein content of the grain is shown in Table 7.3.

Table 7.3.   Influence of nitrogen application on protein content of grain sorghum

Nitrogen rate	Yield	Protein
(lb/acre)	(lb/acre)	(%)
0	3100	6.8
40	4300	6.9
80	5100	7.8
160	6200	10.3
320	6500	12.4

Notice that there was little difference in protein contents at the lower increments of nitrogen applications. These lower rates increased yields so drastically that grain protein was unchanged. The applied nitrogen was diluted over a larger quantity of grain. The higher rates of nitrogen not only increased yield but also increased protein percentage.

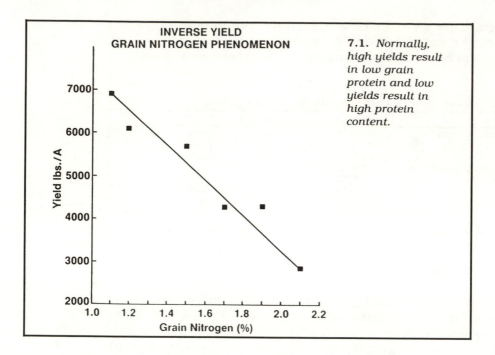

INVERSE YIELD
GRAIN NITROGEN PHENOMENON

7.1. *Normally, high yields result in low grain protein and low yields result in high protein content.*

The inverse relationship between yield and nitrogen percentage of grain sorghum is a well-known phenomenon (Fig. 7.1). The low protein percentage reported from some irrigated grain sorghum has been presumed to be a result of this inverse nitrogen-yield relationship. However, it has also been reported that a linear relationship exists between nitrogen fraction in the grain and the total quantity of water applied just prior to and through the major vegetative growth period. Perhaps the available soil nitrates are moved below the potentially high nutrient absorption zone. As discussed in an earlier section, the high concentration of sorghum roots are in the upper foot, and if the applied water leaches the nitrates below this depth, the nitrate would not be as readily absorbed by the plant. In this case, it would be important that at least a portion of the nitrogen be side-dressed on irrigated sorghums to help insure high-protein grain.

Studies in the Texas High Plains gave the response curves for yields of grain and protein shown in Figure 7.2. Production of both the grain and protein is dependent on the amount of nitrogen applied.

In the studies reported in Figure 7.2, the maximum grain yield was reached at 155 pounds nitrogen, while the maximum protein yield per acre was obtained at about 190 pounds nitrogen per acre. The computed rate of nitrogen for optimum profit is 134 pounds nitrogen per acre.

The optimum rate of nitrogen varies with the relative prices of

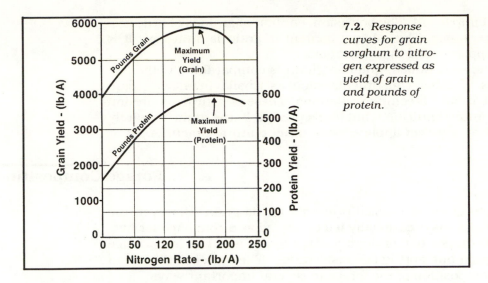

7.2. *Response curves for grain sorghum to nitrogen expressed as yield of grain and pounds of protein.*

grain sorghum and nitrogen. As the price of grain sorghum increases relative to nitrogen costs, the most profitable rate of nitrogen also increases.

Presently, the marketing system does not generally consider the protein content of the grain, even though grain with high-protein content has greater feeding value than grain with low-protein content. Protein content of the grain should be considered as a factor in determining market price. A farmer who feeds grain sorghum or contracts it to a cattle feeder should consider the value of protein in grain sorghum. A cattle feeder might adopt a fertilizer program with higher nitrogen fertilization rates than a farmer who sells grain sorghum on the cash market. The cash grain farmer often subsidizes the cattle feeder by selling high-quality grain at no additional price.

Grain sorghums are high in starch and contain substantial quantities of protein and oil. Starches are easily broken down in the digestive tract of animals into simpler and more easily digested sugars to supply the animals with energy. The oils are primarily fats, which supply even more energy than starches. Protein can also be utilized by animals as an energy source. The main value of grain sorghum in animal rations is to supply energy.

MINERALS

The content of most minerals in the grain can be increased by applying plant nutrients to deficient soils. From a nutritional point of view, phosphorus and calcium are the most important. Phosphorus content of the grain can be almost doubled, but because it is standard practice to supply adequate phosphorus in feed rations, an increase in phosphorus in sorghum sold on the open market is not considered important. The potassium content of grain is of little economic importance. The calcium content of grain is quite low; it is usually supplemented in most rations.

As previously stated, grain sorghum is comparable to other cereal grains in other constituents such as riboflavin, niacin, pantothenic acid, and B-complex vitamins. These constituents are important in animal nutrition, but levels present in the grain are rarely influenced by fertilizer application or other cultural practices.

Forage Composition

Grain sorghum is utilized primarily for its grain. However, the foliage or fodder is occasionally fed or grazed as a roughage for livestock. The composition or quality of the forage is of less importance than the grain but worthy of consideration. The use of fertilizer can alter the composition of the forage in several important ways.

Grain sorghum fodder is fed to livestock for roughage. It is high in carbohydrates and low in protein. When grazed or fed it is almost always supplemented to bring the total ration up to acceptable quality.

PROTEIN AND MINERALS

The composition of sorghum forage is similar to corn forage. At maturity, the ash varies from 5 to 10% on a dry weight basis. Crude fiber varies from 18 to 30%, with crude protein ranging from 5 to 10 percent. The calcium content of the fodder averages about 0.25%, and phosphorus contents from 0.1 to 0.2% are commonly measured. The forage, as noted above, can be quite deficient in protein and phosphorus for livestock feed.

The nutritive value of sorghum forage is greatly affected by the stage of growth. Total protein per acre increases in the sorghum plant until mature, but protein percentage of the stalk decreases rapidly after heading. Protein percentage is high in the immature plant because the cellular structure consists primarily of protein and fiber. The yield as well as the carbohydrate percentage of sorghum increases throughout the life cycle until the plant is mature. The total fiber yield increases as the sorghum plant grows, but the fiber percentage decreases with advancing maturity, and the fiber is fairly digestible until it becomes lignified near maturity.

The quality of forage for livestock is determined by the digestibility, concentration of nutrients, and palatability. Whether fertilizer use influences these factors will determine whether fertilizer increases forage quality. While recognizing that fertilization and a host of factors may influence the quality of forages, the single factor having the greatest effect is stage of growth or maturity.

If grain is harvested early in its growth, forage will be higher in quality than when grain is harvested later. Sorghum plants left after harvest are often utilized for cattle, but quality normally will be quite low.

Nitrogen applications do influence to a very large extent the protein content of the stalk. The protein content of sorghum leaves can be increased two to three times by nitrogen applications to deficient soils. Even though the protein percentage decreases with maturity, the concentration at maturity is a function of the nitrogen in the soil. The protein content is, therefore, easily influenced by fertilizer practices.

Phosphorus fertilization has little influence upon phosphorus content of mature forage, but additional potassium greatly increases potassium concentration of the forage if grown on potassium-deficient soils. Potassium is not a constituent of plant tissue but occurs only in the cell sap. Plants have the ability to accumulate more potassium than they need for maximum growth, and this additional potassium is not generally needed in animal nutrition.

ACCUMULATION OF PRUSSIC ACID AND NITRATES

Grain sorghum plants accumulate both prussic acid (HCN) and nitrates (NO_3^-) under certain conditions. Conditions are more favorable for prussic acid formation when the plant is under stress, and growth is drastically reduced by drought, frost, or other conditions.

The accumulation of prussic acid is primarily in the leaves. Whether a given variety accumulates prussic acid is a genetically controlled factor, while the amount of prussic acid accumulation is influenced by the environment. Nitrogen fertilization tends to increase the amount of prussic acid in the plant. When phosphorus is deficient, prussic acid seems to accumulate. A balanced fertilization program is important to keep prussic acid levels low. Prussic acid is produced as an intermediate stage between the conversion of nitrates into amino acids if the situation is unfavorable for protein synthesis.

Prussic acid levels are highest in young plants, young branches, and tillers. Maximum levels occur at about the 8-leaf stage and decrease until maturity. As forage is cut and dried, the prussic acid content decreases.

Sorghums have been found to accumulate nitrates under certain climatic conditions, with moisture stress being the principal factor. Nitrates accumulate when they continue to be absorbed by the roots but conversion to protein precursors is halted. Most rapid accumulation proceeds when plants are losing turgor during the day and regaining it at night. Also, nitrates tend to accumulate during cloudy weather. Stems accumulate more nitrates than leaves, with the greatest concentration in the lower portion of the stems.

Nitrates tend to increase as nitrogen fertilization increases, while phosphorus and potassium fertilization tends to reduce nitrate accumulation.

Cattle deaths resulting from excess nitrates in corn have been referred to as death from "cornstalk disease." The name is derived from the fact that the animals eat mainly the stalk at this stage of growth. The lower part of the stalk is higher in nitrates than the upper portion. Nitrate poisoning risks can be greatly reduced by feeding a high-energy, low-nitrate supplement to the cattle grazing on the sorghum stalks.

In general, grain sorghum forage contains more prussic acid and nitrates than other sorghum types.

The prussic acid and nitrate contents of grain sorghum are of no real concern to the producer of grain from sorghum since they are rarely grazed by animals except as the residue after harvest. After harvest, toxic levels of these compounds are seldom found in the forage unless regrowth has commenced and then the plant is killed by a frost or freeze. Occasionally, grain sorghum is grazed when the crop is failing. In that situation, the grower should be cognizant of the possible poisoning to livestock. Under proper precautions, grain sorghums can be safely grazed. Toxic levels of both nitrates and prussic acid have been established and are reported in Tables 7.4 and 7.5. Both of these compounds are easily detected and contents can be analyzed in many commercial and university laboratories. Quick tests are available but must be properly interpreted.

Table 7.4. Levels and degree of toxicity of prussic acid in grain sorghum

Level	Relative degree of toxicity
(ppm dry wt, basis)	
0–250	Very low (safe to pasture)
250–500	Low (safe to pasture)
500–750	Medium (doubtful to pasture)
750–1000	High (dangerous to pasture)
Over 1000	Very high (very dangerous to pasture)

Table 7.5. Levels and degree of toxicity of nitrates in grain sorghum

Nitrate-nitrogen content	Toxicity
(ppm dry basis)	
0–1000	Safe under all feeding conditions
1000–1500	Safe for all except pregnant animals
1500–4000	Risk of poisoning; should not be more than one-half the ration
Over 4000	Potentially toxic; should not be fed

8

Water Management

GRAIN SORGHUM is grown principally in areas where rainfall is limited. Much of the world's crop is produced in areas where rainfall is less than 25 inches annually. Therefore, water management, whether under dryland or irrigated conditions, is of utmost importance in producing profitable yields.

Water Needs of Grain Sorghum

Grain sorghum is known as a drought-tolerant plant. This ability to tolerate or resist drought (as pointed out in Chap. 1) is due principally to (1) sorghum plants rolling of leaves as they wilt, thus reducing transpiration, (2) a waxy covering over the leaf protecting it from drying, (3) an extensive root system permitting it to "forage" over a greater portion of the soil (Fig. 8.1), and (4) a relatively small leaf area in proportion to its roots. Sorghum roots will normally

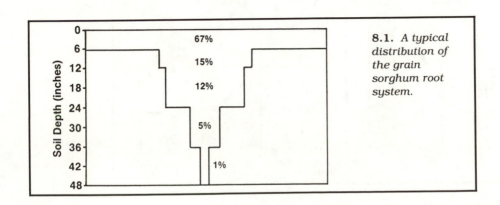

8.1. *A typical distribution of the grain sorghum root system.*

penetrate 3–5 feet if the soil is sufficiently moist and deep. Table 1.1 in the introduction shows that grain sorghum requires less moisture per pound of plant tissue than most other crops. Even though it is called a drought-tolerant crop, grain sorghum will need and utilize relatively large quantities of water if a high yield is to be produced.

Water use by grain sorghum plants will depend partly on the amount of water available. If moisture is adequate, daily water use will usually follow the general pattern given in Figure 8.2. The seed start by imbibing moisture. After germination and emergence, water use is quite low, ranging from 0.05 to 0.1 inch per day during the first 2–4 weeks. When the plant reaches the 6- to 7-leaf stage at about 4 weeks, water consumption begins to increase sharply. It will use 0.1–0.2 inch per day for about 10 days and reach slightly over 0.3 inch during the booting stage. Peak water use may be as high as 0.4 inch per day during the booting and early bloom stages. At this point, the rate of consumption decreases to just less than 0.3 inch per day at 60–70 days. The rate of consumption continues to decrease to the point of physiological maturity, when the consumption is only 0.1 inch per day. After this point is reached, the plant will continue to absorb water and use less than 0.1 inch per day until the plant is destroyed. The moisture absorbed after maturity is of no value to plant growth.

Water use by grain sorghum will depend not only on growth of the crop but also on climatic conditions. In much of the sorghum-producing area, conditions are such that evaporation rates are high and considerable soil moisture is lost.

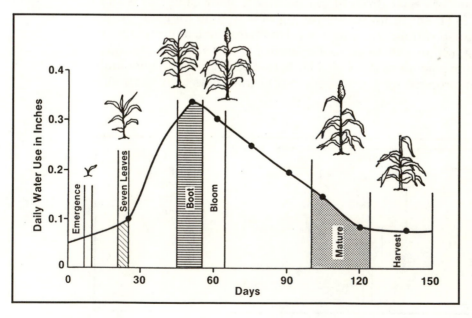

8.2. *Daily use of water by grain sorghum from planting to maturity.*

The stages of booting, blooming, and grain formation compose the *critical water-use period.* For maximum yield, either under dryland or irrigation, adequate moisture needs to be available during these stages of growth. In dryland areas, plant so that the sorghum is booting during the periods when soil moisture is normally apt to be at its highest level. When irrigating, keep soil above 50% field capacity in the top 18 inches.

Highest water-use efficiency by grain sorghum will result when adequate water is available to the plant throughout the season (Fig. 8.3). The crop should not be allowed to go into moisture stress if possible. It has been proven that plants with continuous adequate moisture will outyield plants that are allowed to suffer for moisture. Fortunately, the grain sorghum plant has the ability to withstand periods of moisture stress, but yields are decreased if stress persists too long. It does hurt sorghum to wait for water.

Moderately high yields of grain sorghum will usually require at

8.3. *The most critical time for moisture is at preboot and flowering stages. (Courtesy Leon New, Texas Agricultural Extension Service)*

least 20 inches of usable water during the season. This will vary depending on evaporation, variety, and cultural practices. The amount can come from that stored in the soil prior to planting, rainfall, and supplemental irrigation if available.

Water use by grain sorghum is variable. In cool seasons with high humidity, it is lower than in a hot, dry season. Frequency of irrigation, wind conditions, and loss of water by deep percolation and runoff also affect water use by grain sorghum. The total use is also determined by yield.

Table 8.1 gives an estimate of the water required to produce a certain yield. This table is based on the assumption that sorghums will extract soil moisture to a depth of 4 feet, 10 plants are required per hundredweight of yield anticipated, row spacings must be narrower as plant populations and yields increase, seeding rates are on the basis of 15,000 seed per pound, and 50% of seed planted will survive.

Table 8.1. Estimated water requirements and cultural practices needed for various yield levels

Yield	Water required	Row spacing	Plant population	Seeding rate
(lb/acre)	(in.)	(in.)	(plants/acre)	(lb/acre)
1800	10	40	18,000	2.4
3300	13	40	33,000	4.4
4600	16	20–28*	46,000	6.0
6500	20	20–28*	65,000	8.7
7200	24	10–14	72,000	9.6
7800	28	10–14	78,000	10.0

Note: Adapted from Oklahoma State Univ. Fact Sheet 2216, 1973.
*Or the equivalent of two rows on 40-inch beds.

Table 8.2. Estimated grain sorghum yield based on stored soil moisture and total moisture received (either rainfall or irrigation) on a silt loam soil

Total moisture received*	Depth of penetration (ft)			
	1	2	3	4
(in.)				
8	1800	2800	3700	4600
10	2800	3700	4600	5600
12	3700	4600	5600	6500
14	4600	5600	6500	6900
16	5600	6500	6900	7200
18	6500	6900	7200	7500
20	6900	7200	7500	7800
22	7200	7500	7800	. . .
24	7500	7800	. . .	. . .

Note: Based upon water requirement in Table 8.1 and assuming 2 inches of available soil moisture per 1 foot of wetted soil.
*Refers to both irrigation and rainfall received during the growing season and is not total for the year. Actual yields will vary depending on distribution of rainfall. These figures assume good distribution.

A grower can estimate potential yield on the basis of moisture in the profile at time of planting and moisture normally received in a season. First, estimate the depth of moisture stored in a field. Second, determine the average rainfall for your area. Use the figures in Table 8.2 to estimate the yield. If irrigation water is applied, the figures given should still hold. For example, if a silt loam is preirrigated and wetted to a depth of 4 feet and receives an additional 16 inches of water (rainfall and irrigation), yields should be in the 7200-pound range. If yield is less, check to see what is limiting yields, such as a lack of fertilizer, weed problems, or some other factor.

Water Conservation for Dryland

Conservation of water is usually the most important practice to consider in much of the grain sorghum area. This is accomplished by several means. The most often used practice is to fallow the land for at least 4–6 months prior to planting. Fallowing usually refers to leaving the land without a growing crop and destroying any new plant growth that might appear. This results in no utilization of the moisture that falls, thus permitting it to accumulate in the soil profile. Some of the moisture will be lost by evaporation, but after a dry layer of soil has formed on the surface, evaporation losses are minimized. Most soils in the grain sorghum area can accumulate from 1 to 2.5 inches of moisture in the top foot of the profile. Stored soil moisture is vitally important in dryland sorghum production (Table 8.2).

Another method of conserving moisture is to leave the residue from crops on the surface or only partially incorporated. Wheat or grain sorghum stubble is often left on the surface for this purpose. This is called "stubble mulching." It should decrease losses from evaporation and increase the rate of infiltration.

Wheat-grain sorghum rotations are often used as a means of permitting land to lay fallow and accumulate moisture. After wheat is harvested in July or August, the land will be left without vegetation until planting time the following spring. This will usually mean 7–8 months for moisture to accumulate. After grain sorghum, the land will lie fallow for another 7–8 months prior to seeding wheat, thus allowing accumulation of moisture for the wheat.

Moisture can be conserved by listing and planting on the contour. Fields with considerable slope should be terraced, with rows planted parallel to the terraces. Not only does more of the rainfall infiltrate the soil, but soil erosion also is reduced. Strip cropping is practiced in some areas to conserve snow and reduce wind damage.

Irrigation of Grain Sorghum

WHETHER TO IRRIGATE

Even though grain sorghum is a drought-tolerant crop, it responds well to irrigation (Fig. 8.4). It has been proven that high yields can be obtained under irrigation and that water-use efficiencies for grain sorghum are higher than for many other plants.

Several items need to be considered before deciding whether to develop an irrigation system. The first question would be availability and quality of the water. The next question would be whether the additional moisture will produce sufficient extra yield to pay for the amortized cost of the system.

The additional yield that can be obtained from irrigation will depend upon the yield level that is obtained with normal rainfall. In west Texas, yields of dryland grain sorghum will normally average 1500–2000 pounds per acre. With irrigation, yields will average from 5000 to 7000 pounds per acre with four to five irrigations. In more humid areas where only one or two supplemental irrigations may be needed for top yields, an extra 1000–3000 pounds grain per acre are often obtained with irrigation. Because of variation in rainfall and growing conditions, it is impossible to list the increases that can be expected from irrigation in an area. In situations where irrigation water is limited, alternate-row irrigation will help increase total water use efficiency (Fig. 8.5). Local agriculturalists usually have data developed showing anticipated increases from irrigation for various soil types and land-use capabilities.

8.4. *An application of irrigation water prior to planting places moisture throughout the profile for future use by the crop. (Courtesy High Plains Research Foundation)*

8.5. *Alternate row irrigation. (Courtesy High Plains Research Foundation)*

Irrigation alone will not always lead to increased yield unless other factors responsible for higher yields are controlled. Extra fertilizer, insecticides, and herbicides may also be needed.

Costs of developing an irrigation system would also need to be determined. Local conditions vary considerably; hence no definite costs can be given. If a field is sufficiently uniform in slope and water is readily available, dryland might be converted to irrigated land at relatively low cost; however, if land is undulating and has to be leveled and if water is deep, costs could be quite high. Sprinkler irrigation systems may be used instead of leveling the land.

Several items of cost need to be considered. Fixed costs include water permits (if any, such as in water districts), wells, power unit, distribution of water such as ditches or pipe (underground or surface), land preparation such as leveling, clearing, or need for subsurface drainage, and equipment for moving the irrigation system. Additional items can be listed under operation costs and would include fuel for power, labor, and extra inputs such as seed, fertilizer, chemicals, and harvesting. Costs of maintenance of the power unit, pump, water distribution system, and the like would also need to be considered. A properly designed system can reduce operation and maintenance costs.

Before putting in an irrigation system, check on the quality of available water. All irrigation water contains salt; some waters contain very little while others contain excessive quantities. When total salts are too high or if certain undesirable types of salts are proportionately high, the water could be detrimental to plant growth and eventually result in a greatly reduced yield potential due to accumulation of salts in the soil. Analyses of irrigation water in an area are usually available. However, it is wise to obtain a sample of the water to be used and have it tested.

METHOD OF APPLICATION

Two general methods of applying irrigation water are surface irrigation and subirrigation. Surface irrigation is presently used on the largest number of acres. Subirrigation is practiced only on limited acreage. Only the surface method will be discussed.

Surface irrigation is accomplished by furrow, flood, or sprinkler system. Furrow is the predominant method used (Fig. 8.6), with sprinkler next, followed by flood irrigation. Furrow and flood systems are used mainly on medium- to fine-textured soils, while sprinklers are used mainly on coarse-textured soil. Increased use of sprinklers is being made on medium- to fine-textured soils.

If it is determined that it would be profitable to install an irrigation system, the method to use will be determined by soil type (texture, depth, slope), quantity of water available, climatic conditions (wind, length of season), crops to be grown, availability of labor, and cost.

8.6. *Furrow irrigation is the principal method of applying water in most of the irrigated grain sorghum areas.*

One particular factor, soil texture, may alone determine the system to be used. A sandy soil would normally have to be sprinkler irrigated since it would be impossible to use surface irrigation. The topography of the land may also be sufficiently rolling that the leveling cost would be prohibitive and a sprinkler system would have to be used.

Small flows or quantities of water may be more efficiently used with a sprinkler system. Excessive winds and resultant loss of water by evaporation would suggest surface irrigation. Grain sorghum can be satisfactorily irrigated by either sprinkler or surface irrigation.

Labor may influence the method selected. Automated sprinkler systems utilize less labor than surface irrigation or sprinkler systems that are moved manually.

Cost would obviously be a big factor. Sprinkler systems usually require a higher per acre investment than surface irrigation systems.

In furrow or flood irrigation, the soil is used as a means for conveying and distributing the water over the field; it is the most common method of irrigation. Water is applied in the furrows between rows of plants. As it runs down the row, part of it moves into the soil.

In row irrigation, a surface irrigation ditch or an irrigation pipe (either aboveground or underground) is used to move water to the field. Siphon tubes transfer water from an open irrigation ditch into the row. If irrigation pipe is used to transport water to the field, a "gated" pipe with outlets transfers the water to the row. If the siphon tubes or pipe outlets are properly sized to fit the intake rate of the soil and other factors that influence the need for moisture, this results in a reasonably uniform and efficient method of irrigation. It is more expensive than sprinkler systems from a time and labor standpoint.

Sprinkler irrigation systems also have limitations. The greatest disadvantage is that the initial investment is high. Energy requirements for the operation of the system are also quite high; hence the amount of power needed to run the sprinkler systems would be higher than for surface systems. Sprinkler systems are normally used on soils that would be classified as sands or loamy sands. They occasionally will be used on the sandy loam soils and in some cases even on loam or clay loam soils. On the finer textured soils, sprinkler systems have to be adapted for the lower water infiltration rates.

SPRINKLER SYSTEM

Several types of sprinkler systems are used. They may be either permanent (solid set) or portable. The permanent systems may operate through buried pipe or through aluminum pipe aboveground. The solid-set type is usually used on high-profit crops such as vegetables and fruits and is seldom used for grain sorghum.

Portable systems can be classed as hand-moved, tow-moving, lateral or side-wheel rolling, and automatic continuous moving. The hand-moved system has a low initial cost but high labor requirements. Alternative systems substitute some mechanical means of moving for the higher labor requirement. Investment costs increase since the equipment is easier to move and requires less labor.

The tow-moving systems are on wheels with sprinklers attached to the irrigation pipe, which serve as axles for the wheels. A tractor or pickup is used as the source of power to tow the system to a new setting. In the lateral or side-wheel rolling system, a small power unit is attached to the system described above. The power unit puts torque on the pipe, causing it to roll. This is usually done intermittently when it becomes necessary to move the system.

The mechanical systems that move automatically and continuously are becoming more widespread because of a need to reduce labor. There are both circular (pivot-point) and lateral moving systems (Fig. 8.7). The circular type rotates around a pivot-point anchored in the center of the field and normally makes a complete

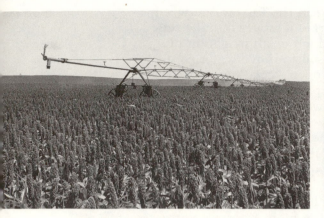 

8.7. *Two types of automatic sprinkler systems: (left) self-propelled and moves around a pivot point in a circular pattern; (right) moves across a field in a rectangular pattern. (Courtesy B & J Industries)*

circle in 2–4 days. Lateral moving automatic systems (similar to the one described in the previous paragraph) are used to some extent. These move in a straight line across a field. Costs of the various sprinkler systems will depend on the complexity of the system. Cost of a hand-moved system is low, whereas a continuous automatic system would be high.

FERTILIZING THROUGH A SPRINKLER SYSTEM

Applying fertilizer through a sprinkler system, "fertigation," is gaining in popularity. It has several advantages. The crop can be "spoon-fed" whereby nutrients can be applied at the time the demand is greatest. Losses due to leaching are minimal. Labor costs may be less than with conventional methods. Efficiency of use may be increased, resulting in lower fertilization rates.

There are also disadvantages. Distribution of plant nutrients will be only as good as the water distribution patterns, which are usually imperfect. Certain fertilizer materials cannot be applied through a sprinkler. Higher cost materials may need to be used. Certain nutrients, particularly phosphorus, may be held at the surface of the soil and not be effectively utilized.

Nitrogen is the principal nutrient applied through sprinklers. It is water-soluble and, therefore, easy to apply in this manner. Nitrogen moves readily into the soil. The most convenient nitrogen fertilizer to apply is the ammonium nitrate-urea solution. It is commonly referred to as 28 or 32% nitrogen solution. Fertilizers with free ammonia should not be used in waters with moderate to high levels of calcium. These include anhydrous ammonia and certain nitrogen solutions that contain free ammonia, such as aqua ammonia.

To determine the rate of nitrogen to apply in a system, first determine the total need for the crop and divide this into several applications based on the normal growth pattern of grain sorghum. Take an example of a need for 120 pounds nitrogen per acre. A portion of this, perhaps 30 pounds, should be applied preplant or in the first irrigation. This would leave 90 pounds to be applied. If four additional irrigations are to be made, apply 10 pounds during the first irrigation, 30 pounds in the second irrigation, 40 pounds in the third irrigation, and 10 pounds in the last irrigation. If the system is a circular type, estimate the number of times the system will cover the field and apply the quantity needed during each rotation. Remember that the highest nitrogen needs are from early booting to the period when the grain is filling. If a rain replaces one of the applications of water, double up on the next.

Applying phosphorus through a sprinkler is risky. If the calcium content of the water is too high, certain reactions occur that may cause a precipitate to form and be deposited in the system. Most phosphorus applied through sprinklers is a solution of ammonium phosphate. Never apply phosphatic fertilizers through sprinkler systems prior to checking with fertilizer dealers or knowledgeable agronomists. Another significant point on phosphorus is that it can be satisfactorily applied to the soil preplant without danger of loss due to leaching. It would be of little benefit to apply phosphorus through a sprinkler.

Potassium, sulfur, and the micronutrients can also be applied through a sprinkler system. Details on sources and quantity to apply can be obtained from most fertilizer companies or university agronomists. Remember that excessive amounts of certain micronutrients may be toxic. Never use iron salts in a sprinkler system (even though iron chelates can be).

Pumps and other metering devices are commercially available for injecting fertilizer into irrigation systems.

DESIGN OF THE IRRIGATION SYSTEM

The proper design of an irrigation system is important when considering irrigation of grain sorghum. Items to consider will include water infiltration rate of the soil; moisture storage capacity; other crops to be grown; source, amount, and availability of water; topography of the soil; and quality of the irrigation water.

The system should be designed, whether for sprinkler or row irrigation, so that the application rate will match soil infiltration capacity, allowing the added water to move into the soil but not run off the surface. Excess water will be wasted and decrease efficiencies. The capacity of the system should be sufficiently large to meet the

8.8. *Irrigation systems need to be properly designed by engineers to be sure pumps and motors are properly placed and are of the right size and capacity. The use of underground pipe and water outlets may save labor. (Courtesy High Plains Research Foundation)*

peak moisture demands of grain sorghum. No more water should be applied at one time than the soil will hold. Excessive irrigation results in loss of water by runoff or excessively downward movement out of the root zone. In addition, some plant nutrients may be leached out of the root zone by excessive watering.

The design of an irrigation system is an engineering job (Fig. 8.8). If it is decided that irrigation is desirable for grain sorghum, the initial contact should be with an irrigation engineer to look at all aspects that will influence the use of the irrigation water. A properly designed irrigation system can mean extra profit in the form of higher yields or savings in time, labor, and equipment.

Irrigating the Crop

TIME TO IRRIGATE

The most obvious time to irrigate is when moisture is needed. Highest yields have been obtained when soil moisture is kept at a level of 50% field capacity or higher. This is particularly true from the time the plant is at the 6- to 8-leaf stage when it is starting its peak growth period and until it has reached physiological maturity or late-dough stage.

Many farmers using irrigation in low-rainfall areas will need to apply a preplant irrigation to provide moisture for planting. Subsequent irrigations will depend on the adequacy of water. If water is plentiful at a reasonable cost, it will usually be applied four more times during the season (unless rainfall comes during the growing season). The first irrigation after the plants are growing is usually at the 6- to 8-leaf stage. It is at this stage of growth and on through the soft-dough stage that moisture requirements are greatest.

Where water is limited, irrigation practices will vary. Most producers will apply a preplant irrigation and then one more at about the boot stage. If there is sufficient water for a second irrigation during the growing season, an ideal time would usually be at the bloom stage or slightly later.

Inadequate moisture for grain sorghum early in the season is much less critical than later. Grain sorghum's drought-tolerant ability comes prior to booting. If water stress occurs during the early-bloom to late-dough stage, yields will be decreased, depending on the extent of the stress or lack of moisture.

With emphasis on water conservation in irrigated areas, attention is being turned to techniques of more efficient water use. Preplant applications may be light just prior to or after planting, with subsequent light applications as needed. In sandy soils, this practice is being used more widely.

Most soils should be irrigated when available moisture declines to about 50%; methods are available for determining this. Most farmers that check will do so by the "feel" method. This is done by using a spade or soil probe to obtain a sample of soil from the root zone. The feel or appearance of the soil will indicate the moisture content.

FREQUENCY OF IRRIGATION

The capacity of the soil to hold water and the rate at which water moves into a soil will determine how often to irrigate. These soil characteristics will also determine the total water that might be available throughout a season and consequently the yield potential.

The ability of a soil to hold water will depend on physical properties of the soil, its texture, and the degree of aggregation (Fig. 8.9). The inches of water that a soil will hold and the amount available to a plant for various textures of soil are listed in Table 8.3. Note that sands will hold less water per foot than other textures, but they will also "turn loose" a larger proportion of the moisture. Clays will hold up to 3–4 inches per foot, but less than 2 inches will be available for plant use.

The rate at which water will move into a soil, called the water infiltration rate, will determine the amount of water that can be applied per hour without any loss due to runoff. It would also greatly influence the design of the irrigation system. Table 8.4 shows the water infiltration rates for various soil textural classes.

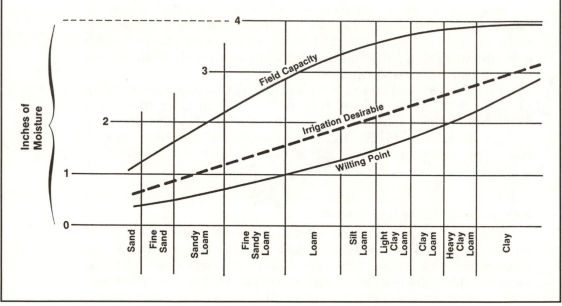

8.9. *Water-holding characteristics of a soil. The difference between the two solid curves is the moisture that is available for plant growth. For example, a fine sand would hold up to about 1 inch, while a silt loam would hold over 2 inches.*

Table 8.3. Approximate available water–storage capacity of soils

Textural class	Total storage	Available water
	(in./ft)	*(in./ft)*
Sands (coarse)	1.0–1.5	0.6–0.8
Sandy loams (coarse to medium)	2.0–2.5	1.0–1.5
Silts and loams (medium)	3.5–4.0	1.6–2.0
Clay loams (medium to fine)	4.0–4.5	2.0–2.5
Clays (fine)	4.5–5.0	1.6–2.0

Table 8.4.   Infiltration rates of various soil textures

Texture	Good physical condition	Poor physical condition
Coarse sand	2.0–3.0	1.2–1.6
Fine sand	1.8–2.0	0.8–1.2
Sandy loams	1.2–1.8	0.5–1.0
Silts and loams	1.0–1.2	0.3–0.4
Clay loams	0.5–1.0	0–0.3
Clays	0.2–0.5	0–0.2

Note: Rates are in inches per hour.

The infiltration rate of a soil will usually be higher initially, then it will slow down within a short time. The latter is the rate that would be considered as the infiltration rate. Normal infiltration rates for soils with good structure as well as rates for soils with poor physical condition are given in Table 8.4. This decrease in infiltration rate is more evident in the fine-textured soils because of surface sealing. Water movement after the initial entry is controlled by movement of water through the profile. The physical condition of the soil and its structure will influence infiltration. If soil structure and aggregation are poor, infiltration rates will be considerably lower.

QUANTITY OF WATER TO APPLY

The quantity of water to apply for irrigation will depend on the amount of water in the soil and that which the soil can hold. The quantities listed in Table 8.3 under total storage are an indication of the quantity of water a soil will hold at field capacity.

A silt loam soil would hold approximately 3–4 inches in the top foot of soil. The available water that can be extracted would usually not exceed 2 inches per foot due to the attraction of the soil particles for water. However, most irrigated grain sorghum producers do not let moisture get that low prior to irrigating. As a result, about 4 inches would normally be needed in the silt loam soil to bring it to field capacity. In most silt loam soils in the irrigated grain sorghum–producing areas, growers will usually apply from 3 to 5 inches of water per application.

QUALITY OF IRRIGATION WATER

Where irrigation water qualities are poor, the salinity problem that arises needs to be given some attention and an effort made to keep the salts from accumulating in the root zone. Overirrigation may be needed periodically so that salts will be leached below the root zone and, therefore, have little effect on the subsequent production of the grain sorghum.

9

Diagnosing Production Problems

GRAIN SORGHUM PRODUCERS occasionally experience problems of inadequate or improper growth of their crop. Diagnosing the cause is important for successful and profitable production. In some cases there may be a marked difference in growth or appearance from one part of the field to the other or between two fields. Diagnosing the problem may be easy, but quite often it is difficult. The diagnosing of crop problems is emerging as a new science, even though growers have been doing this to the best of their abilities for years. Recently, more knowledge utilizing scientific principles has become available in diagnostic work. All the knowledge available must often be used, because visual symptoms can be similar for several different conditions. For example, some disease organisms cause plants to malfunction, resulting in appearances similar to some nutrient deficiencies. It may require close scrutiny and some orderly, organized detection procedures to ascertain the real problem.

Identifying Pest Infestations

For a key to identifying insect and disease problems, see Table 9.1.

INSECTS

Insect damage can vary from the quite obvious to hidden symptoms. If an insect such as the earworm has riddled the leaves and is on the leaves, the problem is obvious and positive diagnosis can

Table 9.1. **Key to sorghum insect and disease problems**

| Symptoms | Damaged by or due to | | |
	Insects	Diseases	Other conditions
Poor germination and poor stand	Wireworm, false wireworm, seed corn maggot, southern corn rootworm, white grub, ant	Fungal seed rot	Directly over ammonia bands, cool weather, dry soil
Poor seedling vigor	Corn leaf aphid	Fungal root rot	
Wilting and dying	Chinch bug, sorghum greenbug	Bacterial leaf blight, fungal leaf diseases, *Periconia* root rot	Mechanical damage, dry soil
Discoloration of leaves	Sorghum greenbug	Bacterial leaf diseases, fungal leaf diseases, maize dwarf mosaic	Nutrient deficiencies, chemical damage
Holes in leaves and whorls, riddled eaten tissue	Corn earworm, grasshopper		Hail damage
Holes in stalks, abscissed layers	Corn earworm, lesser cornstalk borer, southwestern cornborer, grasshopper, armyworm	Fungal leaf diseases	
Damaged heads (early grain)	Corn earworm, midge, webworm, grasshopper, armyworm, chinch bug	Covered kernel smut, loose kernel smut, head smut	Bird damage
Damaged heads (late grain)	Midge, webworm, chinch bug	Covered kernel smut, head smut	Bird damage
Lodged plants (before grain formation)	Southern corn rootworm, white grub	*Rhizoctonia* stalk rot	Wet soils early with subsequent drought
Lodged plants (after maturity)	Lesser cornstalk borer, southwestern corn borer	Charcoal rot, *Fusarium* stalk rot, *Colletotrichum*	Wet soils early with subsequent drought

readily be given. However, if the damage is similar to that inflicted by the greenbug, such as a discoloration of the leaves, the diagnosis is more difficult.

The insects attacking sorghums can be divided into four principal groups. The soil-infesting insects attacking the *seed and seedlings* are usually wireworm, false wireworm, seed-corn maggot, southern corn root worm, white grub, and ants. Insects that most often attack *leaves and stems* (Figs. 9.1, 9.2) are corn earworm, lesser cornstalk borer, southwestern corn borer, grasshoppers, and the several types of armyworms. Sucking insects that usually feed

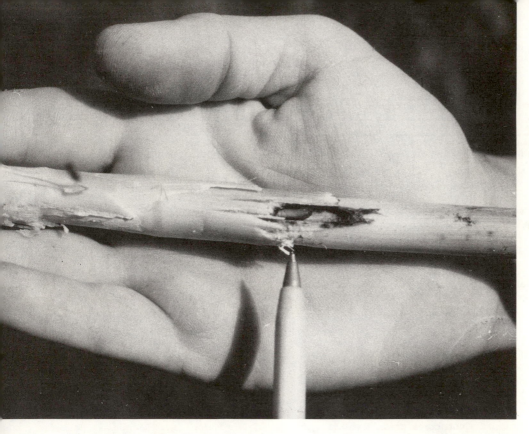

9.1. *Damage caused by the southwestern corn borer.*

9.2. *A stem borer entered the grain sorghum stalk at the internode (left). The inside of the stalk (right) shows the effect of the borer.*

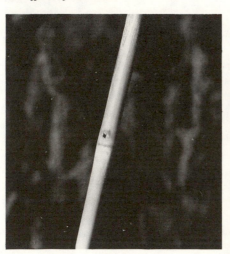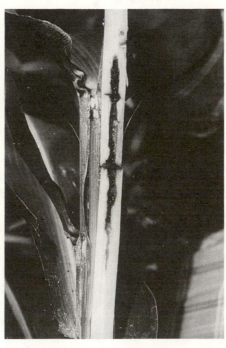

9.3. A. *A severe infestation of sorghum greenbugs causes sorghum leaves to turn purplish, with the midrib tending to remain green.* **B.** *Some varieties have resistance to greenbugs (right) contrasted to susceptibility (left).*

9.4. *Natural enemies of greenbugs are ladybugs and parasitic wasps. Both a ladybug larvae and pupae are shown on the left with the remains of greenbugs. On the right are greenbug mummies after being killed by parasitic wasps.*

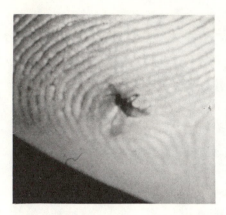 

9.5. *An adult sorghum midge (left) and a comparison of a midge-resistant (large head) and a midge-suscepti-ble variety (small head). (Courtesy Stan Coppock, Oklahoma State University)*

on the *cell sap* are sorghum greenbug (Figs. 9.3, 9.4), corn leaf aphid, and chinch bug. Insects attacking the *head and kernels* include false chinch bug, rice stinkbug, sorghum midge (Fig. 9.5), sorghum webworm, and corn earworm.

To determine whether sufficient insects are present for treatment, insect counts in grain sorghum are taken on a per plant or plant part basis depending upon the feeding habits of the insect involved. Counts should be taken from a representative cross section of the field. At least 25–50 plants should be checked and the number of insects reported as the number per 100 heads, leaves, whorls, or plants. When checking sorghums, remember that insects are most commonly found in the whorl early in the season and in the head later in the year. Check these carefully. The head should be bent over and thoroughly inspected. Then check the ground below to see if insects have fallen out without being noticed. For most insects, threshold levels have been established that would indicate that control measures are desirable.

DISEASES

Diseases are generally harder to diagnose than insects primarily because they are caused by organisms that are so much smaller than insects that they are hard to detect. Disease organisms are rarely seen by the naked eye and may require considerable magnification. Diseases are most often diagnosed by the various symptoms and consequences that are recognized as the results of the action of the disease organism. These include wilting; dieback; root and stem rots; damping-off; cankers; witches broom; stunting; unthriftiness;

poor yields; shriveled kernels; blighted, spotted, discolored, or deformed foliage; seed decay; and numerous other manifestations of an abnormal condition.

Those that rot the *seed* or kill the seedlings are fungal seed rots—*Fusarium, Aspergillus, Rhizopus, Rhizoctonia,* and *Penicillium*—and fungal root rots—*Pythium, Fusarium moniliform,* and *Penicillium oxalicum.* Those that attack the *leaves* are bacterial leaf diseases—bacterial stripe and bacterial spot—and fungal leaf diseases—rough spot, anthracnose (Fig. 9.6), leaf blight, zonate leaf spot, gray leaf spot, target spot, sooty stripe and rust, and maize dwarf mosaic virus (Fig. 9.7). Diseases that attack the *head* are smuts—covered kernel, loose kernel, and head smuts. The major *root* disease is *Periconia* root rot, often referred to as milo disease. *Stalk* rots include weak-neck, charcoal rot (Fig. 9.8), *Fusarium* stalk rot (Fig. 9.9), *Colletotrichum* stalk rot, and *Rhizoctonia* stalk rot.

Chemical control of many disease organisms is effective. Pesticides and suggestions for use for disease control change rapidly; therefore, no specific recommendations will be made here. Agricultural chemical dealers and local agricultural extension personnel have available up-to-date recommendations for disease control.

Cultural practices other than chemical application are often recommended for disease control. Resistant hybrids, residue management, and crop rotations are frequently effective.

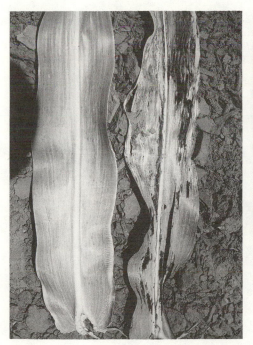

9.6. *Visual symptoms of anthracnose disease of sorghum leaf (left) and impact on two different varieties in the field (right).*

9.7. *Maize dwarf mosaic virus (MDMV) on a plant as well as on sorghum heads. Note darker grain and stalks on heads on right side.*

9.8. *Internal damage of plant tissue caused by charcoal rot (left) and effect on plants in the field (right).*

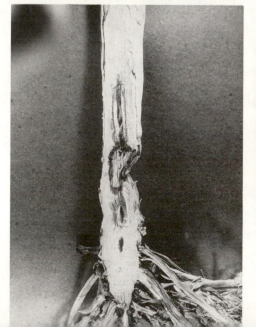

9.9. *Internal damage of the stalk by Fusarium stalk rot (left) and an example of damage to the entire field (right). (Courtesy Ervin Williams, Oklahoma State University)*

BIRD DAMAGE

Birds consuming the grain can cause considerable damage and lower harvest yields. Most bird damage is done at the soft-dough stage, when the sugar content of the kernel is high. Birds start eating the grain at the top of the head, working downward (Fig. 9.10). Missing kernels with some glumes intact along with broken spikelets are some characteristics of bird damage. Bird damage can be minimized by the use of bird-resistant varieties.

9.10. *Sorghum head in which the grain has been partially consumed by birds.*

Identifying Nutritional Problems

A key to identifying nutritional deficiencies in sorghum is found in Table 9.2.

Table 9.2. Key to nutrient deficiency of sorghums

Symptom	Element deficient
Stunted plant	All deficiencies
Loss of green color	All deficiencies
Color changes in lower leaves	
Yellow discoloration from tip backward in form of V along midrib	Nitrogen
Brown discoloration or "firing" along outer margin from tip to base	Potassium
Purpling and browning from tip backward; in young plants entire plant shows general purpling	Phosphorus
Color changes in upper leaves	
Emerging leaves show yellow to white bleached bands in lower part of leaf	Zinc
Young leaves show interveinal chlorosis along entire length of leaf; leaves may eventually turn white	Iron
Uniform yellowing of upper leaves	Sulfur

Source: Adapted and modified from Krantz and Melsted.

USE OF PLANT ANALYSIS

Plant analysis as a diagnostic guide in assessing nutrient needs is an old concept but is playing an increasing role in recent years because of the economic need for obtaining top yields and because new techniques have made the concept easier to accomplish and more accurate.

Plant analyses are done in the field as "quick tests" or in the laboratory (Fig. 9.11). Laboratory tests are more accurate, but the

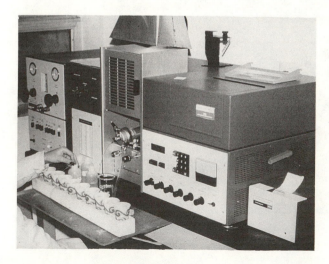

9.11. *Many laboratories are fully automated and computerized for rapid recommendations. Tests have meaning only if the results are calibrated with carefully conducted field experiments.*

field tests are more rapid. Field tests normally measure the unassimilated soluble nutrients of the cell sap. Laboratory analyses measure both the elements that have been incorporated into plant compounds and those that are present as soluble constituents of the plant sap.

Plant analyses can be an aid in (1) determining the sufficiency or deficiency of nutrients, (2) locating areas on the verge of becoming deficient in a particular element, (3) evaluating fertilizer programs, and (4) suggesting additional and more complete tests to identify the problem.

Some limitations to plant analyses are: (1) They do not always indicate the cause of the problem. (2) By the time plant analyses reveal the problem, it is often too late to correct on the existing crop. (3) Interpretations are difficult because critical concentration levels have not been established for all conditions. Because the concentration of plant parts varies from one growth stage to another, critical concentrations of each nutrient must be established for each plant part at each stage of growth.

Critical nutrient levels for sorghums are given in Table 9.3. Critical levels are defined as the minimum concentration required to maintain adequate production. Factors affecting growth such as soil moisture, temperature, and sunlight will influence the critical levels. Values somewhat lower than those given may be applicable before measurable growth and yield responses can be demonstrated from the application of the nutrient.

In spite of the limitations, plant analyses are useful guides, supplement the soil test in assessing the nutrient status of the plant, and help in making fertilizer recommendations.

To sample for plant analysis, collect from 15 to 20 leaves at random from the field or trouble spot. Select the first fully developed leaf

Table 9.3. Critical nutrient concentrations for grain sorghum

Element	Concentration
	(%)
Nitrogen	2.40
Phosphorus	0.20
Potassium	2.20
Calcium	0.40
Magnesium	0.25
	(ppm)
Iron	25
Manganese	15
Boron	10
Copper	5
Zinc	15
Molybdenum	0.2

Note: Based upon the leaf immediately below the flag leaf during booting and flowering.

below the whorl. If taken at flowering, select the leaf just below the flag leaf (second leaf). For samples to be sent to a laboratory, dust off plant material (but do not wash) and air dry for 1 day. Place each sample in a large envelope (not plastic or polyethylene bags) and mail to a laboratory for analysis. Include questionnaire or information requested by the laboratory.

USE OF FIELD TESTS

On-the-spot field tests are easily and rapidly done. Without leaving the field, the test results can be compared with the appearance of the crop. One advantage of field testing is that it requires a first-hand look and causes the grower to take a close look at the crop. The first prerequisite of interpreting field test results is to observe other possible limiting factors in crop production.

The optimum time for field testing is when the plant is under the greatest stress, usually at midseason when flowering and seed setting begin. The best use of field testing, however, is repeated testing throughout the season. To get the most from field tests, test regularly every 2–3 weeks from the time sorghum is knee-high through the period when the seed is in the soft-dough stage. If testing is done only once, the best time is during booting or early flowering.

For field tests to be properly interpreted, an understanding of plant nutrition and physiology is most helpful. In the hands of an expert, field tests are valuable tools, but erroneous conclusions can be drawn without sufficient knowledge of nutrition and physiology. Nevertheless, laypersons use them to good advantage if they are cautious and recognize the limitations.

Never tissue test plants that are suffering from moisture stress or from water-logged soils. For general observations, sample only plants that are healthy and growing vigorously. Do not sample plants that are suffering from diseases or insect damage because the tests may not be reliable. If an element is deficient, other nutrient tests may not be reliable because other nutrients may accumulate that would give wrong interpretations.

The *nitrate* test on sorghums is the most reliable of the major elements tested. The supply of inorganic nitrogen in the plant is directly related to the nitrate content of the cell sap. Any decrease in the supply of nitrogen to the plant roots is quickly reflected in the cell sap. In sorghums, the highest concentration is usually found at the base of the stem, and the concentration becomes progressively lower toward the top of the plant. In testing for nitrates, a vertical section of the whole plant or the midrib of the leaf can be used. Absence of nitrates can be detected several days prior to visual yellowing of the plant (Fig. 9.12). Testing should be done in midmorning, and subsequent testing should be done at about the same time each day. There is little need to test for nitrates until the plant is at

9.12. *Nitrogen deficiency symptoms are first noted by a light green appearance of the entire foliage. The lighter yellow strip through the field above received no nitrogen, contrasted to applied nitrogen on each side.*

least knee-high because most young sorghum plants contain adequate nitrates. Reagents, equipment, and directions for testing for nitrate can be obtained from several commercial laboratories.

If sorghum plants are deficient in nitrates early in the season, nitrogen can be side-dressed and yield depression caused by inadequate nitrogen can be avoided. The nitrate test on sorghum signifies only whether there is enough nitrate present when the test is made; it does not assure that there will continue to be enough until the sorghum is mature.

The following rating is suggested for nitrates in sorghum:

1. None—no nitrate detectable in sorghum plant.
2. Low—nitrates detectable at base of stalk only.
3. Medium—nitrates present in lower leaves and upper stalk only.
4. High—nitrates present in midribs of upper leaves.

Whenever sufficient *phosphorus* is present for rapid growth, inorganic phosphates accumulate in the cell sap. Field tests measure only these readily soluble inorganic phosphates.

Tissue tests for phosphorus are used to evaluate the fertilizer program for the current crop and assess the needs for future years rather than correct any deficiency for the current crop being tested.

When tissue testing for phosphorus, use midribs from the youngest fully mature leaves (such as the second leaf from the top). Phosphorus tests at the blooming stage give the most accurate measurement.

All the *potassium* taken up by the plant exists in the cell sap and is in the soluble fraction. Select the youngest, most fully mature leaf for testing. Potassium levels are frequently high in young sorghum, but these levels may drop by flowering. Check levels early in the season (knee-high) and again at flowering. If potassium levels drop, the potassium fertilizer rate may have been too low.

There are some field tissue testing methods for *micronutrients*, but not nearly enough research information is available to establish reliability with them. For the micronutrients, the quantitative laboratory plant analysis procedures should be used.

USE OF SOIL TESTS

The use of soil tests in determining fertilizer need has become an accepted practice. It is an indispensable tool for that purpose, but it is also of great value in diagnosing production problems. Soil tests for available nutrients are often needed to supplement information from plant analyses and plant appearance. Soil tests from problem areas compared with tests from normal sites can often be of great value in diagnostic work.

In addition to testing for nutrient deficiencies, soil tests are also being used as a diagnostic tool to determine the presence of other problems. Residual or carryover residue from pesticides, especially herbicides, can be determined. Presence of nematodes can be ascertained by analyzing the soil.

When using soil tests for any purpose, sample collection is vitally important. Remember, the soil test can be no better than the soil sample. *Take samples properly.*

DEFICIENCY SYMPTOMS

Another important tool in diagnosing nutrient deficiencies is learning to recognize the characteristic visual symptoms, frequently called "hunger signs," that a plant develops when deficient in one or more nutrients. In some cases these symptoms afford a better understanding of the nutrient relationships between soil and plant than can be obtained from detailed chemical analysis of the soil or the plant.

Nutrient deficiency symptoms are not always easy to interpret because so many factors can cause the plant to exhibit characteristics similar to those expressed when nutrients are lacking. These include disease and insect infestations, drought, temperature effects (both cold and heat), excess soil moisture, herbicide residue damage, wind or sand damage, and other mechanical damage. Some effects, such as certain diseases or cold temperatures, are often actual nutrient deficiencies within the plant, even though the soil supply is adequate. Environmental factors may result in the inability of the plant to absorb and/or translocate a particular nutrient. For example, a sharp drop in the temperature in the spring will result in a reddish purpling of young leaves, which is symptomatic of phosphorus deficiency. The reddish purpling of the leaves is due to an interruption of the plant's metabolic processes.

Some deficiency symptoms may be quite similar for each of several nutrients. Interveinal chlorosis, for example, can be caused by a deficiency of magnesium and certain micronutrients. To determine which element causes a specific symptom, the soil and its chemical characteristics have to be considered. For example, calcareous (high pH) soils usually contain adequate magnesium, and micronutrient availability may be low; hence, if the interveinal chlorosis is present in sorghum on calcareous soils, it would probably be due to a micronutrient deficiency such as iron, manganese, or copper. If the soil were acid, the reverse would be true. To determine which micronutrient is deficient, the liquid form of the micronutrient can be sprayed on selective plants to determine which one alleviates the deficiency.

Multiple deficiencies become extremely difficult to interpret because symptoms tend to blend when more than one element is lacking. For example, when nitrogen is low, an iron-deficient sorghum plant will not be green enough to manifest the visual sharp contrast in green and chlorotic tissue characteristically associated with iron-deficiency chlorosis in its early stages.

In spite of the difficulties and the caution that must be exercised, deficiency symptoms can be used to help diagnose field problems and in some instances help evaluate the effectiveness of the fertilizer treatment.

Fortunately, sorghum, with its wide expanse of broad leaves, is a better indicator of changes in the supply of available nutrients than most other crops. It is like corn in this regard, and its characteristic symptoms resemble those observed on corn.

A brief description of the common deficiency symptoms described for sorghums is listed below (Fig. 9.13).

Nitrogen deficiency is characterized by stunted and spindly growth in the young grain sorghum plant. The foliage will be light green in color instead of dark green. In older plants, the tips of the lower leaves become yellow with the yellowing moving up the leaf midrib in a V-shaped pattern as the deficiency persists. Later, the yellow tissue dies. This type of firing can also be caused indirectly by hot dry weather and a lack of moisture.

In very young sorghum plants, a *phosphorus deficiency* is expressed as a reddish purpling on leaves. If the deficiency persists, the plants are stunted and become very dark green. Finally, the leaf tips become necrotic as the tissue dies. Small heads of grain, late maturity, and unthrifty plants may indicate a phosphorus deficiency.

The first symptoms of *potassium deficiency* are a shortening of

9.13. *Specific deficiency symptoms include (A) yellowing (chlorosis) down the midrib of the lower leaves for nitrogen, (B) browning (necrosis) of the outer edge of the lower leaves for potassium, (C) general yellowing of upper leaves and interveinal chlorosis for iron, and (D) emerging leaves showing a broad band of bleached, yellow tissue on each side of the midrib for zinc.*

the internodes. A more severe deficiency produces a bronze to yellow discoloration along the edges of the lower leaves. The discoloration is continuous all around the leaf margin. The outer edge of the discoloration becomes necrotic (dies) as the yellowing works inward toward the midrib.

Deficiency symptoms for *sulfur* are stunted growth, delayed maturity, and a general yellowing of the foliage. In some cases, an interveinal yellowing may occur. Sulfur deficiencies are normally expressed first in the younger leaves. Deficiency symptoms due to a lack of sulfur seldom are seen on soils with a pH of 6.0 or above.

Zinc deficiency is characterized by a broad band of bleached tissue on each side of the midrib, giving it a very striking striped appearance. It usually begins on the lower half of the leaf as the leaf is unfolding.

Sorghum is probably the best indicator crop to detect *iron deficiency* (Fig. 9.14). As with zinc deficiencies, interveinal chlorosis or striping exists, but with iron deficiency it extends the full length of the leaves. The chlorosis starts with the youngest or uppermost leaves. In cases of severe deficiency the chlorophyll fades out completely, leaving the plants white, and eventually they may die.

9.14. *The severity of iron deficiency chlorosis will vary over a field. Note the two darker green rows on each side of the picture where two foliar applications corrected the deficiency.*

Identifying Other Factors Affecting Growth

Many other factors may limit crop growth and final yields and need to be considered.

SOIL CHARACTERISTICS

The most common soil characteristics needed for high levels of production in addition to high levels of fertility are soil physical properties such as good aeration, good but not excessive drainage, and good water infiltration rates.

Compacted Soils. One of the most common physical problems of soil is compacted zones in the soil. They inhibit root proliferation and growth. These compacted zones may be either natural or tillage induced. The tillage-induced zones are referred to as "plowpans."

Some soils are so clayey that the claypan inhibits root growth. Some soils become sealed and tend to be "cemented" by the planter opener if moisture content is too high at planting. Upon drying, the soil becomes so hard that plant roots cannot penetrate it. Regardless of the cause, reduced rooting depth can greatly decrease yield by causing premature soil moisture stress. In addition, uptake of nutrients is restricted. In diagnosing production problems, the first step should be to make an examination of the physical condition of the soil.

Saline Soils. Soil salinity and/or alkalinity may often be a problem. Excessive salts of calcium and magnesium in the soil may inhibit water and/or nutrient uptake and reduce growth. Excessive salts of sodium in a soil may lower productivity because of the poor physical condition that exists. These are called sodic (alkali) soils. The excessive salts may be related to high water tables, in which case the only permanent treatment is by drainage. For sodic soils where no high water table is present, the area may be treated by applications of agricultural gypsum or sulfur.

HERBICIDES

Abnormal and poor plant growth has often resulted from improper use of herbicides (Figs. 9.15, 9.16). Herbicide damage to sorghums has been noted especially on coarse-textured soils and soils low in organic matter. Abnormalities such as chlorosis or stem and head distortions develop that may resemble nutrient deficiency symptoms.

9.15. *An application of 2, 4-D sprayed on a grain sorghum field to control broadleaf weeds resulted in damage to roots and subsequent lodging.*

9.16. *Lack of grain formation in the heads as a result of spraying with 2, 4-D at the wrong growth stage.*

FERTILIZER PLACEMENT

Injury to sorghum can result from improper fertilizer placement. A high concentration of nitrogen and potassium may cause seedling damage if placed too close to the seed. Ammonia improperly placed can cause young sorghum seedlings to die. Ammonia burn causes the plant to lose its turgidity and turn yellow. Root pruning while applying anhydrous ammonia or during cultivation can cause plants to wilt temporarily.

LACK OF MOISTURE

When sorghum plants are subjected to severe moisture stress, they turn dark green or have a bluish color. They wilt, and the upper leaves roll. A mild nitrogen deficiency is sometimes confused with moisture stress because the lower leaves become chlorotic and die (Fig. 9.17). Remember that a yellowing of the lower leaves is not a symptom of drought conditions.

9.17. *If lack of moisture is severe enough for tissue to die, it will first show in the younger leaves, with dead tissue turning almost white. "Firing" of lower leaves is often blamed on a lack of moisture, when, in most cases, it is a lack of nitrogen.*

WEATHER HAZARDS

Weather-related problems can often occur, such as hail damage, too high or too low temperature, and high winds. Damage by hail can be confused with insect infestation unless the plants are carefully observed. Insects leave tell-tale residues. Hail most often leaves some broken plants (Fig. 9.18). After maturity, sorghum seed can germinate if very humid weather persists for several days. This condition is commonly referred to as sprouting in the head (Fig. 9.19).

9.18. *Hail damage of grain sorghum. Note ragged leaves; in some cases, only the midrib is left. Grain was also damaged.*

9.19. *Sprouting of grain while it is still in the head is often a problem in high-humidity/high-rainfall areas.*

Steps in Diagnosing Field Problems

1. Obtain all available information from grower.
   (a) Soil test reports, which will give the general fertility levels of the field.
   (b) Past cropping history, which will help in determining fertilizer requirement. Be on the lookout for crops with heavy nutrient removals. Previous yields will be of significant value.
   (c) Past fertilizer treatments. Look for residual or carryover of nutrients as well as deficient levels. Don't forget to check on lime or other soil amendments.
   (d) Current management. Find out what chemicals were used for weed and insect control and what tillage and seeding practices were used. Remember that date of planting and variety may be important.
2. Field observation.
   (a) Look for deficiency symptoms, insects, disease, and weed infestations.
   (b) Make plant population counts.
   (c) Study the roots and depth of rooting.
   (d) Examine the soil for compacted zones and excessive salts.
3. Run field tests. Compare normal plants with abnormal ones if available.
4. Collect samples for plant and soil analyses.

With all the available information, attempt to diagnose the problem. Somewhere along the line a negative diagnosis may have been erroneous. If a positive diagnosis cannot be made, don't forget a multiple complication. Often a group of factors are responsible for poor growth. The important task is to try to decide which factor is the most limiting. Often the only solution is elimination of the problem by trial and error.

10

Harvesting, Marketing, and Utilization

THE GRAIN SORGHUM CROP is not produced until it is harvested, stored, and sold. The more there is to harvest, the more important it becomes to pay attention to details when harvesting and marketing.

Losses of 5–10% before or during harvesting is not too uncommon. With a yield of 2000 pounds per acre, this is only 100–200 pounds, but with a yield of 8000 pounds per acre, the loss would be 400–800 pounds, almost the entire profit in some cases. So increase profit by watching the details of harvesting.

Harvesting

Most grain sorghum is threshed from standing stalks with a combine (Fig. 10.1). Prior to 1940, 10% of the grain sorghum planted was combined; 90% was "bundle feed," which was usually cut, stacked in the field, and hauled in later for use (Fig. 10.2). Grain with up to 25% moisture can be harvested. Drying equipment is needed at this moisture level for storage of the grain. Without drying equipment, most grain is harvested at approximately 13–15% moisture.

10.1. *Combining grain sorghum.*

10.2. *"Bundle feed" stacked in the field.*

Early harvest may be one key to highest profits. When grain sorghum has reached maturity, its dry weight, grain quality, and yield are at its highest level. After that point, both yield and grain quality will decrease with time.

Early harvest will prevent losses due to stalk breakage, hail, and similar hazards. In areas of high humidity later in the fall, sorghums might even gain moisture if the grain is left too long in the field. Acreage harvested per machine may also be increased with early harvest, thus possibly reducing cost.

Potential losses increase as grain harvest is delayed (Fig. 10.3).

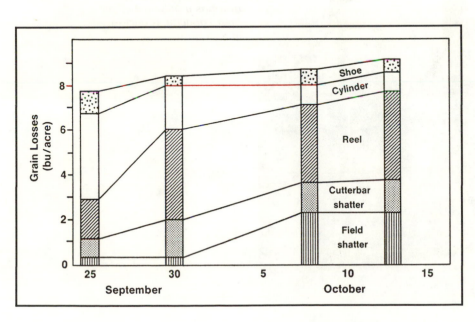

10.3. *Relationship between date of harvest and harvesting losses and the combine part responsible for the loss.*

The increase in losses is mainly from field shattering due to the effects of the combine reel. Losses from early harvest come principally from the cylinder. The effects of the shoe and cutterbar on losses are small. Be sure the reel is properly adjusted to avoid losses at this source.

The sickle bar and cylinder should be properly adjusted to avoid losses. Grain is easily damaged while being combined, particularly if the grain is dry enough for storage. High-moisture grain will usually need a higher cylinder speed, which may also cause increased cracking of the grain. Compared to wheat, cylinder speed for grain sorghum is normally one-half as fast.

If the grain sorghum is lodged severely at harvest, it may be necessary to use a pickup attachment on the combine; most combine manufacturers have such attachments. Lodging is usually caused by damage from insects such as the stalk borer or due to disease damage such as stalk rot. One of the advantages of an early harvest is that losses due to stalk lodging will be less.

Early harvest often results in high-moisture grain, which means that drying will be needed. It may also necessitate recleaning or extra effort at cleaning if weeds are a problem.

Artificial drying of grain sorghum on the farm is gaining in popularity (Fig. 10.4). It offers the obvious advantages that go with early

10.4. *On-the-farm drying facilities permit earlier harvesting of grain. This unit has a 365-bushel-per-hour capacity. (Courtesy of Kan-Sun)*

harvest—mainly lower field losses. The main question is, Does it pay? It is a matter of comparing the cost of drying to the amount saved by early harvest. Cost of drying will vary over the sorghum-producing areas, so find out what it would cost to dry in your area and then estimate how much you can save and decide whether to have a drying system on the farm.

High-moisture grain may be dried in the field or after it gets to the elevator. Drying equipment used in the field is either a portable batch dryer or a continuous flow system. Either one appears to be satisfactory. A dryer should be adjusted for temperature and properly timed. If temperature is too high, the quality of the grain may be reduced; if too low, the cost of drying per unit will increase.

If grain sorghum is dried in the field or in storage on the farm, it may be desirable to have a moisture tester. It could save money, because overdrying means that less weight is taken to the elevator. There usually is no premium for grain sorghum that is below the normal moisture level. Overdrying also costs extra for the fuel used. Moisture testers are relatively inexpensive.

With the advent of feedlots in the grain sorghum–growing area, some high-moisture grain is being harvested and used. It is usually fed within a relatively short time after it is stored. Cattle feeders are showing more interest in storing and feeding high-moisture grain.

Artificial drying methods can also be used before harvest. Chemicals can be used to speed drying and hasten the harvesting of grain sorghum. The chemical partially kills the plant, which permits a more rapid drying of the grain and vegetative parts. Desiccants have not been used to any great extent in grain sorghum production since conditions in most of the grain sorghum–producing area are such that grain will normally dry without the use of desiccants. They may prove to be of value to the seed producer as well as commercial producers in humid areas. In addition to the standard desiccant type of chemicals, nitrogen solutions containing urea and ammonium nitrate have often been used for this purpose.

Storage

Grain sorghum should be stored between 11 and 13% moisture. With proper aeration, it can be safely stored up to 15–16% moisture. Most grain elevators have drying systems and will take grain over the 13% moisture level but will usually adjust the price downward because of the higher moisture content.

On-the-farm storage has been used only to a small extent in grain sorghum areas (Fig. 10.5). Grain can be satisfactorily stored on the farm. This practice will probably increase so that the producer can take advantage of seasonal price variations.

10.5. *On-the-farm storage is gaining in popularity in some areas. (Courtesy Butler Company)*

The important point to remember in grain storage, whether on the farm or in commercial storage, is to keep losses at a minimum. Losses in both quality and quantity can occur. Quality losses come about in several ways. Molds can cause smutty grain. Heat damage can cause discolored grain and deterioration of the germ. Sprouting can occur. Insects can also reduce quality.

The two principal factors of concern in storing grain sorghum are moisture content (and relative humidity) and the temperature at which it is stored. These two are in turn influenced by time. High-moisture grain can be stored at low temperature for short periods without damage. But if the temperature is raised or the grain is stored for a longer time, problems can develop.

Molds will develop where the temperature is above 75°F and moisture is above 15% for a period of several months. The grain can be kept for a longer period without problems from mold development by decreasing the moisture to 13%. Heat damage is evidently related to moisture and damage from mold.

Insects are a major hazard to stored grain. They consume grain, which results in losses. They also cause heating and spoiling and thereby a reduction in grade. Insect damage is also influenced by temperature and moisture. There are certain optimum ranges in which insects will be active. Chemical fumigants provide good control of insects. Proper safety precautions in the use of fumigants need to be observed.

Trash in grain sorghum may cause trouble in storage because it increases the danger of heating and spoiling. Trash is usually in the form of stems, stalks, weeds, and leaves as well as cracked kernels. Properly adjusted combines will help prevent trash problems.

Properly constructed bins are necessary for proper grain storage. A tight structure is essential—one that will keep out moisture, rodents, and insects.

If grain storage or drying is planned, contact engineering representatives of storage building companies or manufacturers of drying equipment. They can provide information on proper storage conditions for moisture, temperature, insect control, and the length of time grain can be stored.

The use of silos to store high-moisture grain sorghum is receiving some attention. Grain sorghum can be satisfactorily stored if properly handled. Silos may be used more in areas of high humidity where drying in the field in the fall may be a problem.

Marketing

Marketing of grain sorghum is changing because patterns of utilization are changing. Exports are trending downward, while domestic use has been increasing. Years ago, most of the grain sorghum was exported from the principal growing areas to cattle markets outside the area or into the export market. Grain was sold to the local elevator (Fig. 10.6), which held it in storage, or it was moved to the terminal elevators. It was then processed into feed or shipped to the markets for feeding or export.

10.6. *The initial marketing and storage of grain sorghum is handled principally by commercial country elevators. (Courtesy Butler Company)*

With the recent trends toward increased local consumption by feedlots in the producing area, a large percentage of the grain sorghum is utilized in areas where it is produced. The local elevator usually receives the grain during harvest. Quite often, it is sold directly to the elevator at that time. However, the trend in marketing now is toward holding the grain to take advantage of normal seasonal price increases and government programs.

Additional changes have taken place in marketing grain sorghum, most of which have permitted producers to receive a higher price for grain. Expanded markets include direct sales to the greatly expanded livestock feeding industry. Grain sorghum is being used in increasing quantities in livestock and poultry rations.

Standards for marketing grain sorghum have been established in the United States by the Department of Agriculture. Most of it is sold as No. 2 grain sorghum. The standards used to establish various grades are described in Table 10.1.

Table 10.1. Grades and grade requirements for grain sorghums

Grade	Minimum test weight/ bushel	Moisture*	Total	Heat- damaged kernels	Broken kernels,* foreign materials, other grains
1	57	13	2	0.2	4
2	55	14	5	0.5	8
3†	53	15	10	1.0	12
4	51	18	15	2.0	15

SAMPLE: Sample grades shall be grain sorghum that does not meet the requirements of any of the grades from No. 1 to No. 4, inclusive; contains stones; is musty, sour, heating, or badly weathered; has any commercially objectional foreign odor except of smut; or is otherwise of distinctly low quality.

Source: USDA Official Grain Standards of the United States. SRA-AMS-177, 1964 (rev. 1988).

*Values are a maximum for that grade.

†Grain sorghum that is distinctly discolored shall not be graded higher than No. 3.

The damaged kernels listed in Table 10.1 can include several types of damage (Fig. 10.7). *Heat damage* is probably one of the most important and prevalent types. *Black germ damage* results from going out of condition in storage. *Mold and ground damage* results from the grain head having been on the ground or mold taking place in the head due to high rainfall or humidity. This should not be confused with discolorations that occur from normal weathering. *Insect damage* includes kernels that have been bored by insects, usually weevils, but no longer have the insects in the grain. If the insects were present, it would be included in the special grade of *weevily grain*. *Sprout damage* includes grain that has sprouted. *Badly weathered* grain results from grain staying in the field too long.

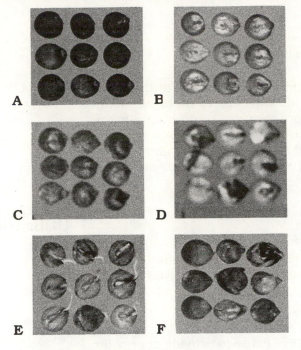

10.7. *Kernel damage influences the value of grain sorghum and the grade it will receive. Illustrations of factors considered: (A) heat damage, (B) black germ damage, (C) mold and ground damage, (D) insect damage, (E) sprout damage, (F) badly weathered. (Courtesy Oklahoma State University)*

Broken kernels, foreign material, and other grains will influence the grade as shown in Table 10.1. This will include small kernels, broken kernels, and other matter that will pass through 0.2 cm triangular hole sieve. It will also include any foreign matter that is bigger. Other grains include almost all grains other than corn and small grains. It also includes the nongrain sorghums: sweet sorghum, sorghum-sudangrass hybrids, johnsongrass, and similar type crops (Fig. 10.8).

Johnsongrass in grain sorghum is a more serious mixture than in forage sorghums. It is a troublesome, hard-to-control weed and is declared a noxious weed in most states. Johnsongrass-sorghum crosses, such as sorghum almum and similar johnsongrass deriva-

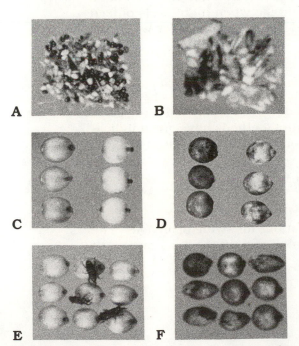

10.8. *Six other grading factors are used in classifying grain sorghums: (A) dockage (foreign material), (B) cracked kernels and other grains, (C) bright grain, (D) weevily, (E) smutty, (F) nongrain sorghums (sudan, johnsongrass, etc.). (Courtesy Oklahoma State University)*

tives, must be considered serious mixtures in seed and grain.

In addition to the grades listed in Table 10.1, there are some special grades used for grain sorghum. Smutty grain sorghum includes grain where the kernels are covered with smut spores or contain 20 or more smut masses per 100 grams of grain sorghum. Weevily grain sorghum is grain infested with live weevils or other live insects injurious to grain sorghum.

Utilization

The biggest increase in use of grain sorghum is for livestock feed. It is also exported. A small amount of the crop is used for food and industrial purposes.

BEEF CATTLE

Grain sorghum may be used for beef cattle to provide energy for both maintenance and fattening conditions. Generally, if the grain is processed, no difference should be observed when substituting grain sorghum for any other feed grain. Sorghum grain has satisfactorily composed as high as 98% of fattening rations. There seems to be no restrictive level.

DAIRY CATTLE

Data are limited on the processing of grain sorghum and its effect on milk production. With the steam-flaking processing method, there is a tendency to lower percent butterfat. If high levels of grain are fed to milk cows, butterfat will usually be depressed. Generally, the feeding of grain would make up a small portion of a ration for a dairy cow; therefore, one should observe no difference in substituting grain sorghum for any other feed grain. Sodium bicarbonate is useful in maintaining butterfat percentage when high levels of feed grain, grain sorghum, or corn are fed in the ration and/or when the feed grain has been steam flaked.

SHEEP

Grain sorghum that has been dry rolled (coarsely cracked) is acceptable for sheep because they tend to chew their feed more completely than cattle. The other processing methods have given indications of being slightly less effective for sheep than for fattening cattle.

SWINE

Grain sorghum is an excellent feed for swine and may be substituted for any of the cereal grains in rations for swine of all ages. The substitution rate may vary from a small amount such as one-tenth or less to complete substitution for other cereal grains being used in the ration. The substitution may be made on a unit-for-unit basis. Grain sorghum may be fed to swine in the whole grain form; however, the feeding value would be increased by grinding or rolling to a medium degree, especially for older animals. Steam processing, flaking, and other processing methods have no advantage over grinding sorghum grain for swine.

Processing Grain for Feeding

Processing grain sorghum by passing it between two rollers is relatively common. Dry rolling is sometimes referred to as "cracking" or "crimping." It should result in coarsely cracked grain with a minimum of fine particles. Steam rolling is accomplished by holding the grain in a pressurized chamber for about 1.5 minutes at approximately 1 pound per inch prior to passing the grain between the rollers. After rolling, the flattened flakes weigh approximately 22 pounds per bushel and contain 18–20% moisture. The flaked grain is partially dried by an airlift system that takes the grain to storage. Data on the use of flaked (steam-rolled) grain for cattle and sheep indicate about a 10% increase in efficiency as opposed to dry rolling.

Popping is a process that allows the grain to be inverted in the form of a popped kernel. It is popped in dry air by heating to about 350°C. The popped material is then passed through a roller. Only about one-third of the grain actually pops prior to rolling. The volume of the grain is approximately doubled by the process. Its moisture content is around 7%.

Micronizing is a relatively new process. Infrared heat waves penetrate the grain sorghum to the point of almost causing the grain to pop or invert before passing between two rollers. Gas-fired generators produce the infrared. Microwaves penetrate the grain as it vibrates along the heating surface. The end product is a crimped flake that is low in moisture and over twice the volume of the original grain.

High-moisture storage is gaining in popularity by feeders. Grain is stored at a moisture level of between 25 and 32% in an airtight structure for at least 18 days. There are two methods for preparing high-moisture grain: (1) the grain can be harvested from the field in the high-moisture state and stored, or (2) water can be added to air-dry grain to raise the moisture content to the desired level prior to storing, which is termed reconstitution of grain sorghum. With either method, the grain is generally rolled or ground prior to storage. Generally, high-moisture storage can be considered equal to steam and/or pressure flaking. Scant information would suggest ranking micronizing and popping near flaking and high-moisture storage.

11

The Future of Grain Sorghum

GRAIN SORGHUM has been aptly referred to as the "wonder crop" of semiarid and arid agriculture. It produces grain in areas too hot and too dry for corn production. It is a unique plant, possessing some characteristics that allow it to escape moderate droughts, especially prior to head formation.

Production of grain sorghum has been revolutionized in the past 40 years. Yields have increased dramatically. The development of hybrids in the 1950s coupled with irrigation, fertilizer use, and other improved practices in the 1960s and 1970s resulted in high-yield potentials. In the 1980s, further refinement in specificity of hybrids for certain conditions took place. Consistently high yields and acceptance of the grain by the livestock feeding industry has firmly placed grain sorghum as a competitor in the feed grain industry.

The future for grain sorghum is indeed bright. However, to compete with other grains, improvement in production must continue. Plant breeders must continue their efforts in improving cultivars that are higher yielding and pest tolerant or resistant. Increasing head length and seed size appears promising. Improved quality of amino acid content is necessary.

The future grain sorghum plant should be even shorter, more efficient in regards to moisture and nutrient use, and morphologically shaped to enable higher plant populations.

Fertilizers need to be improved and more efficient placement methods developed. A better understanding of the soil processes responsible for supplying available nutrient forms throughout the growing season is needed. The interaction between sorghum's physiological mechanisms responsible for nutrient uptake and utilization and soil processes needs to be understood.

154

Better fertilization practices, soil acidity and alkalinity control, and water management offer substantial improvement in yields.

Reductions in energy requirements to produce grain sorghums are possible by reducing tillage and cultivation requirements. Further developments in herbicide effectiveness can accomplish these reductions.

The popularity of grain sorghum as a grain crop is expected to increase. The following trends in production seem evident.

1. The hybrids grown will continue to change. The new hybrids will permit higher plant populations. They will have longer heads and larger seed sizes. The newer hybrids will be resistant and/or tolerant to more diseases and insects.

2. Narrower rows will be used to accommodate the higher plant populations.

3. Improved cultural practices and improved hybrids will allow for earlier seeding.

4. Less tillage and cultivation will reduce energy requirements and improve soil physical properties as new herbicides are developed and application methods are improved.

5. Fertilizer use will increase as yields increase and other cultural practices change. More efficient application methods will be devised and employed.

6. Irrigation of sorghums will continue where water is available and sorghum production in humid areas will increase. New hybrids and improved cultural practices will allow for more grain production per unit of water.

7. There will be continued improvement in machinery design. Machinery will be more efficient. Harvest and storage losses will decline.

8. Production of grain sorghum will keep pace with the demand, which is expected to increase as the world need for food increases.

GLOSSARY

Abscission—the natural separation of leaves, flowers, and fruits from the stems of plants by the formation of a special layer of thin-walled cells.

Absorption area (root system)—the root surface area capable of taking in water and nutrients.

Acid soil—a soil with a pH of less than 7.0.

Actinomycetes—a microorganism similar to fungi and bacteria that is active in the decomposition of soil organic matter.

Adventitious structure—a structure arising from a plant part other than the normal.

Aftermath growth—regrowth of forage after harvest for hay or grain.

Alkaline soil—a soil with a pH above 7.0.

Alkali soil—a soil with a pH of 8.5 or higher, usually a result of excessive sodium (soils with an exchangeable sodium percentage above 15), also referred to as sodic soil.

Ammonia (NH$_3$)—a nitrogen-containing compound; a gaseous form of fertilizer nitrogen.

Ammonium (NH$_4^+$)—a form of nitrogen taken up by plants.

Anther—the saclike structure of the stamen in which microspores are produced.

Anthesis—the developmental stage at which anthers rupture and pollen is shed.

Antibiosis—a situation where one organism is harmful to another.

Bifurcated—divided into branches (forked).

Blade—the wide part of a leaf above the petiole.

Blasted heads—heads without seed development.

Boot (sorghum)—the stage of growth just prior to head emergence.

Brace roots—roots developing from basal nodes near the ground level that serve to support the plant.

Buttress roots—the same as brace roots.

Calcareous soil—soil containing free calcium carbonate and in which the concentration exceeds the cation exchange capacity.

Carbon-nitrogen ratio—the ratio of weight or organic carbon to the weight of total N (mineral plus organic forms) in soil or organic material; commonly used to refer to the carbon–organic nitrogen ratio when inorganic N levels are low.

Carotene—a pigment; vitamin A precursor.

Caryopsis—the seed or single grain.

Chelate—an organic molecule that reacts with a metal ion and keeps it in a more readily available form.

Chlorosis—loss of chlorophyll in a plant creating a light green or yellow color.

Compacted soils—soils with a high-bulk density; dense soils.

Cotyledons—embryonic leaves that serve as food-storing organs.

Crop residue—the portion of the plant remaining after harvest.

Crop rotation—a systematic change of one crop to another in a planned sequence on the same land.

Cross-pollinated sorghum—sorghum plants developed by the process in which pollen is transferred from an anther of one flower to the stigma of a different plant.

Culm—the jointed stem of a grass plant; the stool or tiller.

Cytoplasm—the material and components of a cell other than the nucleus.

Deficiency symptom—a visual plant abnormality resulting from an insufficient level of a nutrient.

Desiccant—any chemical used to dry plants, usually in preparation for harvest.

Diagnosis—a study and examination to determine the cause of a plant abnormality.

Dormant—the stage of growth or development in which the rate of metabolic processes is at a minimum.

Double dwarf sorghums—the term used for mutant or genetic recombinations that resulted in shorter varieties than common in the early part of this century; the height class now called combine sorghum.

Dough stage—stage of growth in which seed has formed but has not hardened; kernels are soft or "doughy."

Durra—a class of sorghum found in northeast Africa, the Middle East, and India and first introduced into California in 1874.

Embryo—the minute plant produced as a result of fertilization and development of a zygote in a seed.

Endosperm—food-storage material in a caryopsis that develops as a result of double fertilization.

Ensile—to make or store forage as silage.

Enzyme—a protein that functions as a biological catalyst and regulates cellular functions.

Fallow—land left uncropped for one or more seasons to collect moisture, destroy weeds, and allow decomposition of crop residues.

Fertigation—application of fertilizers through an irrigation system.

Fertilizer—any organic or inorganic material of natural or synthetic origin that is added to a soil to supply one or more elements essential to the growth of plants.

Feterita—a variety of sorghum originating in northeast Africa of the Caudatum group and introduced into the United States in the early 1900s.

Field capacity—the percentage of water in the soil after movement of water downward by gravity (or free drainage) has ceased.

Fixation—*see* Reversion.

Flag leaf—the first leaf below the head.

Florets—the reproductive subunit of a spikelet.

Foundation seed—a class of seed that is the progeny of the breeder's seed and in which genetic purity and identity is maintained.

Fritted trace elements (frits)—sintered silicates containing micronutrients with controlled (relatively slow) release characteristics.

Fungi—living organisms resembling *Actinomycetes* and bacteria responsible for many plant diseases.

Fungicide—any chemical used for controlling fungi.

Furrow irrigation—a system of irrigation in which water is applied in furrows.

Genetic lines—lines of sorghum that are genetically alike.

Germ—a common name for the embryo.

Germination—the sequence of events occurring in a viable seed leading to the growth and development of a new plant from the seed.

Germplasm—that part of the cell, be it cytoplasmic or intranuclear, that controls hereditary characteristics, i.e., those traits passed on to an organism's offspring.

Glume—the bractlike structure at the base of a spikelet.

Hardening—the result of many changes that occur in a plant as it develops resistance to adverse conditions.

Hardpan—an impervious layer in a soil that restricts root penetration as well as movement of air and water.

Hegari—a variety of sorghum introduced into the United States in 1908 and commonly grown for bundle feed.

Herbicide—any chemical used for the control of weeds.

Heterosis—a process whereby a hybrid will display increased vigor and usually increased yield.

Host plant—a plant supporting an insect or disease vector.

Hunger signs—*see* Deficiency symptom.

Hyaline—transparent or translucent.

Hybrid—a cross between parents differing in one or more characteristics; the F_1 generation.

Indicator plants—plants particularly sensitive to some conditions that are used to diagnose production problems.

Infestation—the buildup of a certain population of a pest.

Inflorescence—the arrangement of the flowers on the floral axis; a flower cluster.

Inoculum—an organism used to build resistance or immunity in a plant to that organism.

Internode—the portion of the stem between two consecutive nodes.

Interveinal chlorosis—yellowing between the veins.

Ion—an electrically charged atom with a surplus or deficiency of electrons.

Iron-deficiency chlorosis—the loss of chlorophyll in plant tissue resulting from a deficiency of iron in the plant.

Kafir—a class of sorghum originating in southern Africa and often a component of modern hybrids.

Kaoliang—a group of sorghums from China and Manchuria that possesses good cold tolerance and a subcoat associated with high tannin.

Kernel—the seed or single grain.

Lemma—the outer bract formed at the base of a grass floret.

Lime—calcium carbonate ($CaCO_3$) or calcium-magnesium carbonate ($Ca \cdot MgCO_3$) used to correct soil acidity.

Lister—a tillage tool used to make furrows.

Lodicules—the scalelike structures at the base of the ovary in a grass floret.

Macronutrient—an essential nutrient in a plant in relatively large amounts (usually more than 500 ppm).

Maize—common term used for grain sorghum in some localities; used to denote corn in most countries other than the United States.

Male sterile plants—those plants in which the anthers are void of or fail to dehisce pollen.

Mesocarp—the middle layer of a pericarp.

Micronutrient—an essential nutrient present in plants in relatively small quantities (normally less than 50 ppm).

Milo—a variety of sorghum introduced from northeast Africa in the 1880s and significant to the breeding of improved hybrid sorghums.

Minimum tillage—a system where the number of times a field is tilled is reduced to the minimum required to raise a crop.

Moisture stress—when plant growth is affected by a lack of available soil moisture.

Mulch—a material applied to or left on the surface of the soil for conservation of moisture and/or erosion control.

Mutants—plants resulting from a spontaneous change in the genetic makeup of the cell.

Necrosis—death or decay of plant tissue.

Nematode—microscopic animal, usually worm-shaped, that can be parasitic on plants.

Nitrate (NO$_3$)—a form of nitrogen taken up by plants.

No-till—a farming system where a field is not tilled except at planting time; chemicals are normally used to control weeds.

Nutrients—substances essential to the growth of plants, such as nitrogen, phosphorus, and potassium; an available nutrient is one that can be readily taken up by the plant.

Offshoot—short, horizonal stems that develop from the crown of stems.

Off-types—types of sorghum not conforming to the original; types different from that planted.

Optimum fertilizer rate—one that will give the maximum net profit per acre.

Osmotic pressure—pressure developing in cells resulting from diffusion through a semipermeable membrane separating two solutions.

Outcrossing—a process of producing seed of the same variety but of slightly different types; a cross to an individual not closely related.

Ovary—the basal, generally enlarged part of the pistil in which seeds are formed.

Palea—the innermost, smaller, bractlike structure enclosing a single grass floret.

Peduncle—the top part of a stem that supports a head.

Pericarp—the walls of the ovary at maturity.

Permutation—a rearrangement, such as of the genetic factors, that controls inheritance.

Pesticide—any chemical used to kill pests, e.g., weeds, insects, and fungi.

Photosynthesis—the process whereby solar energy plus carbon and oxygen from carbon dioxide and hydrogen from water are converted into food energy in the form of carbohydrates.

Physiological maturity—a stage of growth when a plant, or the grain of a plant, has reached its maximum dry weight.

Physiology—processes occurring within a plant; collectively, all the processes and phenomena of plant growth and reproduction.

Plant population—the number of plants per unit area growing in a field.

Pollen—the microspore, a male gametophyte; the end product of meiosis in an anther; the male gametes or sperm in plants.

Pollination—the transfer of pollen from an anther to a stigma.

Pollinator—a plant used to produce pollen for crossing.

Predator—an organism that lives or preys on another organism.

Prussic acid (HCN)—a toxin that is produced in sorghum and other plants.

Residual nutrients—essential plant nutrients remaining in the soil after a crop is harvested.

Reversion—the changing of an essential plant nutrient from soluble to less soluble forms as a result of interaction with, or reactions in, the soil; usually refers to the conversion of phosphorus from a soluble (available) form to a less soluble (less available) form.

Rhizome—a horizontal stem growing partly or entirely underground; often thickened and also serving as a storage organ.

Root hair—an extension of the epidermal cell of a young root immediately beyond the root tip.

Roughage—carbonaceous forage used to supply fiber in cattle rations.

Saccharine—sweet; of the nature of sugar.

Safened seed—seed that have been chemically treated as a protection against a herbicide; it allows a herbicide to be used where the safened seed are planted and only the targeted species is affected.

Saline soil—a soil containing excess salts; a soil with an electrical conductivity greater than 4000 mhos/cm.

Secondary roots—brace roots; buttress roots; any roots developing outside the embryo.

Seedbed—the zone of soil in which seed are planted.

Seedling stage—the stage of growth of sorghum from seedling emergence to about the 6-leaf stage.

Seminal root—a root that differentiates in the embryo.

Shallu—a class of sorghum.

Shattercane—an off-type of sorghum.

Sheath—the basal portion of a grass leaf that surrounds the stem.

Silage—forage, usually corn or sorghum, preserved in a moist condition and that results from a partial fermentation in an anaerobic environment in silos.

Soil-buffering capacity—capacity of a soil to resist change, such as a change in soil pH, this capacity coming from clay and organic matter plus compounds such as carbonates and phosphates.

Soil fertility—the status of a soil with respect to its ability to supply the nutrients essential to plant growth.

Soil management—the manipulation of soil factors affecting crop production so that optimum yields can be achieved at lowest cost.

Soil pH—a measurement of acidity or alkalinity expressed as the negative logarithm of the hydrogen ion concentration.

Soil reaction—the acidity or alkalinity expressed in general terms as being either acid or alkaline.

Soil texture—the proportion of sand, silt, or clay in a soil.

Spikelet—a basic morphological unit of grass inflorescences; may contain one or more florets.

Stamen—the part of the flower consisting of the anther in which pollen is produced; a slender filament that holds the anther in a position favorable for pollen dispersal; male reproductive organ.

Stomata—a microscopic opening in leaves through which gas exchange for photosynthesis, respiration, and transpiration occur.

Stubble mulching—leaving residue on the soil surface or only partially incorporated.

Suckering stage—the stage of growth during which tillers or shoots arise from an axillary bud.

Tannins—polyphenols associated with a subcoat or testa layer in the kernel, often effective in reducing bird damage or surface mold but deleterious when reducing protein digestibility.

Tillage—the mechanical manipulation of soil for any purpose; in agriculture it is usually restricted to the modifying of soil conditions for crop production.

Tiller—the culm, stool, or stem of a grass plant.

Tilth—the physical condition of the soil with respect to its fitness for planting or growing a crop.

Toxin—harmful to an organism at a specified concentration.

Transpiration ratio—pounds of water required to produce a pound of plant material.

Variety tolerance—the ability of a variety to tolerate an insect or disease organism without being unduly affected.

Water infiltration rate—the rate (usually in inches per hour) at which water moves into the soil.

Xanthophyll—the natural yellow pigments in plants.

Yield—the production of a crop on a specified area such as pounds per acre.

Yield potential—the maximum yield possible for a certain soil or field when all available knowledge and techniques are ideally applied under a given set of conditions.

Zygote—the cell produced by the union of the male gamete (sperm) with the female gamete (egg); grows into the miniature plant or embryo.

Bibliography

Selected general references on grain sorghum

Bebee, C. N. Sorghums and Millets Biography (April 1967–August 1978). Bibilogr. Lit. Agric. 4. USDA-SEA Tech. Inf. Syst. 1979.

Byrd, A., K. R. Teferteller, and A. Pope. Higher sorghum yields + extra feeding value. Plant Food Rev. Winter, 1960.

Chaffin, W. Sorghums for Grain and Forage. Okla. Ext. Circ. E-478. 1973.

Clark, L. L., and D. T. Rosenow. Off-Type Sorghum Plants. Texas A&M Publ. MP-885. 1968.

Crabb, A. R. The Hybrid Corn Makers: Prophets of Plenty. Rutgers Univ. Press, New Brunswick, N.J. 1947.

Dotzenko, A. O., N. E. Hamburg, G. O. Hinze, and W. H. Leonard. Effects of Stage of Maturity on the Composition of Various Sorghum Silages. Colo. Agric. Exp. Sta. Tech. Bull. 87.

Elrick, G. L., R. C. Long, F. C. Strickler, and A. W. Pauli. Stage of Maturity, Plant Population, and Row Width as Factors Affecting Yield and Chemical Composition of Atlas Forage Sorghum. Kans. Agric. Exp. Sta. Tech. Bull. 139. 1964.

Grain Sorghum Handbook. Kans. Coop. Ext. Serv. 1975.

Grain Sorghum Research in Texas. Texas A&M Univ. PR-2938-2949. 1971.

Krantz, B. A., and S. W. Melsted. Nutrient deficiencies in corn, sorghum, and small grains. In Howard Sprague, ed. Hunger Signs in Crops, 3rd ed., pp. 25–27. David McKay Co., New York. 1964.

Lane, H. C., and H. J. Walker. Mineral Accumulation and Distribution in Grain Sorghum. Tex. Agric. Exp. Sta. Publ. MP-533. 1961.

Locke, L. F., and O. R. Matthews. Cultural Practices for Sorghum and Miscellaneous Field Crops. USDA Circ. 959.

Maunder, A. B. Agronomic and quality advantages for yellow endosperm sorghums. Proc. 26th Annu. Corn Sorghum Res. Conf., pp. 42–53. 1971.

National Research Council. Nutrient Requirements of Beef Cattle. National Academy of Sciences, Washington, D.C. 1984.

Proceedings of the Feed Grains Utilization Symposium. Texas Tech Univ., GSPA Sorghum Promot. Comm. and Texas Cattle Feeders Assoc. Sept. 20, 1984.

Proceedings of the Grain Institute. Texas Tech Univ., Lubbock. 1970.

Ross, W. M., and O. J. Webster. Culture and Use of Grain Sorghum. Agric. Handb. 385. ARS, USDA. 1970.

Shull, G. In A. R. Crabb, The Hybrid Corn Makers: Prophets of Plenty, p. 66. Rutgers Univ. Press, New Brunswick, N.J. 1947.

Stephens, J. C., and Holland, R. F. Cytomplasmic male-sterility for hybrid sorghum seed production. Agron. J. 40:20–23. 1954.

163

Thompson, C. A. Fertilizing Dryland Grain Sorghum on Upland Soils in the 20–26 inch Rainfall Area in Kansas. Kans. Agric. Exp. Sta. Bull. 579. 1972.

Tucker, B. B., and W. F. Bennett. Fertilizer Use on Grain Sorghum in Changing Patterns in Fertilizer Use. Soil Science Society of America, Madison, Wis. 1968.

Webster, O. J., and R. E. Karper. Proc. 26th Annu. Corn Sorghum Res. Conf., pp. 42–53. 1971.

INDEX